Real-Time Multi-Chip Neural Network for Cognitive Systems

RIVER PUBLISHERS SERIES IN CIRCUITS AND SYSTEMS

Indexing: All books published in this series are submitted to the Web of Science Book Citation Index (BkCI), to SCOPUS, to CrossRef and to Google Scholar for evaluation and indexing.

The "River Publishers Series in Circuits & Systems" is a series of comprehensive academic and professional books which focus on theory and applications of Circuit and Systems. This includes analog and digital integrated circuits, memory technologies, system-on-chip and processor design. The series also includes books on electronic design automation and design methodology, as well as computer aided design tools.

Books published in the series include research monographs, edited volumes, handbooks and textbooks. The books provide professionals, researchers, educators, and advanced students in the field with an invaluable insight into the latest research and developments.

Topics covered in the series include, but are by no means restricted to the following:

- Analog Integrated Circuits
- Digital Integrated Circuits
- Data Converters
- Processor Architecures
- System-on-Chip
- Memory Design
- Electronic Design Automation

For a list of other books in this series, visit www.riverpublishers.com

Real-Time Multi-Chip Neural Network for Cognitive Systems

Editors

Amir Zjajo
Delft University of Technology
The Netherlands

Rene van Leuken
Delft University of Technology
The Netherlands

Published 2019 by River Publishers
River Publishers
Alsbjergvej 10, 9260 Gistrup, Denmark
www.riverpublishers.com

Distributed exclusively by Routledge
4 Park Square, Milton Park, Abingdon, Oxon OX14 4RN
605 Third Avenue, New York, NY 10017, USA

Real-Time Multi-Chip Neural Network for Cognitive Systems / by Amir Zjajo, Rene van Leuken.

Routledge is an imprint of the Taylor & Francis Group, an informa business

ISBN 978-87-7022-034-7 (print)

To my son Viggo Alan, and my daughter Emma

Amir Zjajo

Thanks to all my family members, friends and colleagues for their continued support over all these years.

Rene van Leuken

Contents

Preface

This volume presents novel real-time, reconfigurable, multi-chip spiking neural network system architecture based on localized communication, which effectively reduces the communication cost to a linear growth. The system use double floating-point arithmetic for the most biologically accurate cell behavior simulation, and is flexible enough to offer an easy implementation of various neuron network topologies, cell communication schemes, and models and kinds of cells. The system offer a high run-time configurability, which reduces the need for resynthesizing the system. In addition, the simulator features configurable on- and off-chip communication latencies, as well as neuron calculation latencies. All parts of the system are generated automatically based on the neuron interconnection scheme in use. The simulator allows exploration of different system configurations, e.g. the interconnection scheme between the neurons, the intracellular concentration of different chemical compounds.

List of Contributors

Andrei Ardelean, *Delft University of Technology, Delft, The Netherlands*

Jan Christiaanse, *Delft University of Technology, Delft, The Netherlands*

Jaco Hofmann, *Delft University of Technology, Delft, The Netherlands*

Eralp Kolagasioglu, *Delft University of Technology, Delft, The Netherlands*

Rene van Leuken, *Delft University of Technology, Delft, The Netherlands*

Haipeng Lin, *Delft University of Technology, Delft, The Netherlands*

Xuefei You, *Delft University of Technology, Delft, The Netherlands*

He Zhang, *Delft University of Technology, Delft, The Netherlands*

Amir Zjajo, *Delft University of Technology, Delft, The Netherlands*

List of Figures

List of Tables

List of Abbreviations

ADC	Analog-to-Digital Converter
ANN	Artificial Neural Network
ARM	Acorn RISC Machine
AS	Axon+Soma Computational Hardware
ASC	Application Specific Co-processor
ASIC	Application Specific Integrated Circuit
BRAM	Block Memory
CHL	Complete Hidden Layer
CLT	Curve Length Transform
CMOS	Complementary Metal-Oxide-Semiconductor
CPU	Central Processing Unit
DAC	Digital-to-Analog Converter
DEND	Dendrite Computational Hardware
DMA	Direct Memory Access
DSP	Digital Signal Processor
ECG	Electrocardiography
EDP	Energy Delay Product
EPSC	Excitatory Postsynaptic Potential
ExpC	Exponent Calculation Co-processor
ExpC	Exponent Co-processor
FF	Feed Forward
FF	Flip-Flop
FIFO	First In First Out
FIR	Finite Impulse Response
FP	Floating Points
FPGA	Field Programmable Gate Array
GPU	Graphics Processing Unit
HH	Extended Hodgkin–Huxley Model
HH	Hodgkin–Huxley
IC	Integrated Circuit
ICA	Independent Component Analysis

ION	Inferior Olivary Nucleus
ION	Inferior Olivary Nucleus Network
ION	inferior Olive Neurons
LFLI	Layered Full Lateral Inhibition
LNLI	Layered Neighbor Lateral Inhibition
LSM	Liquid-State Machine
LTD	Long Term Depression
LTP	Long Term Potentiation
LTP	Long-Term Plasticity
LUT	Look-Up Table
MIT-BIH	Massachusetts Institute of Technology – Beth Israel Hospital
ML-II	Modified Lead Two
MPR	Multipath Ring
Mul/Div	Multiplication/Division
NOC	Network on Chip
PCA	Principal Component Analysis
PhyC	Physical Cell
PPC	Physical Cells Per Cluster
PSTDP	Pair-based Spike-Timing-Dependent Plasticity
RISC	Reduced Instruction Set Computer
RMS	Root Mean Square
RNDC	Randomly Connected
SimC	Simulated Inferior Olivary Neuron
SNN	Spiking Neural Network
SPI	Serial Peripheral Interface
STDP	Spike Timing-Dependent Plasticity
STP	Short-Term Plasticity
TSTDP	Triplet-based STDP
UDP	User Datagram Protocol
VHDL	Very High Speed Integrated Circuit Hardware Description Language
Vivado HLS	Vivado High Level Synthesis
Vivado HLS	Vivado High Level Synthesizer

1

Introduction

Amir Zjajo and Rene van Leuken

Delft University of Technology, Delft, The Netherlands

1.1 A Real-Time Reconfigurable Multi-Chip Architecture for Large-Scale Biophysically Accurate Neuron Simulation

Continuous neuroscientific progress has gradually led to the realization of mathematical models of the neuron cells and their intricate networks [1, 2], realistic models, which simulate biological behavior with a large level of accuracy, as in the case of spiking neural networks (SNNs) [1, 3]. In SNNs, propagated information is not only encoded by the firing rate of each neuron in the network, as in artificial neural networks (ANNs), e.g., perceptron [4] but also by amplitude, spike-train patterns, and transfer rate. The high level of realism of SNNs and more significant computational and analytic capabilities in comparison with ANNs, however, limit the size of the realized networks. Consequently, the main challenge in building complex and biophysically accurate SNNs is largely posed by the high computational and data transfer demands. In addition, biological neuron networks are characterized by co-localized memory and calculations and execute computations with a high degree of parallelism, for which the conventional von Neumann CPU-based execution is not very well suitable. Due to their inherent high level of parallelism, reconfigurable hardware, such as field-programmable gate arrays (FPGAs), are capable of providing sufficient performance for real-time and even hyper-real-time simulations of these collective and distributed networks. Furthermore, the reconfiguration property of FPGA provides the flexibility to modify the network topologies and the brain models on demand (e.g., Izhikevich model [2, 5, 6], integrate and fire (IaF) model [7] (and its extensions such as the *leaky* IaF, IaF-or-burst [8], *quadratic* IaF [9]), Hodgkin–Huxley (HH) model [10–12], *simplified* Hodgkin–Huxley model [13], and *extended* Hodgkin–Huxley model [14]).

Small-scale special purpose systems, such as ROLLS [15] intended for cortical-like computational modules, implement 256 IaF neurons. In [16], the system containing several tens of thousands of leaky IaF neuron cells are implemented on Virtex-7 FPGA platform. In [17], analog-based Neurogrid system replicates neurons as an electrical system [18]. With 16 Neurocores, i.e., the computation elements of the Neurogrid, the system is able to simulate over 1 million quadratic IaF neurons with billions of synapses. TrueNorth in [19, 20] contains 4096 cores, totaling 1 million programmable digital IaF spiking neurons and 256 million configurable synapses. Although significantly larger number of neurons can be simulated when compared to the FPGA solutions, the Neurogrid and TrueNorth platforms are neither very flexible concerning model changes, nor the neuron behavior can be as easily observed. The models are not analyzed as applications in general, but only as implementations on the particular platform. Additionally, no neuroscientific experiment instances with biological plausibility were considered. Large-scale experiments are performed with a large number of general-purpose processors as in the SpiNNaker project [21] as well, where over 1 million low-power ARM cores are connected by a fast mesh-based interconnect link. Subsequently, the largest SpiNNaker system is able to simulate over 1 billion neuron cells of Izhikevich type [2].

However, for electrochemically accurate neuron modeling, which is a focus of our study, the conductance-based multi-compartment Hodgkin–Huxley model [10] is required. Biophysically accurate models of biological systems, such as the ones using the Hodgkin–Huxley formalism, comprised mostly of a set of computationally challenging differential equations often implementing an oscillatory behavior. If the interconnectivity between oscillating neurons is also modeled (e.g., gap junctions, input integrators, and synapses), the cells become coupled oscillators. Consequently, all neuron states need to be completely updated at each simulation step to retain correct functionality. As a result, cycle-accurate, transient simulator is necessary. The above difficulties in associated HH models and multi-compartmental models with complex connections, in conjunction with biophysically plausible neuron network sizes, pose significant challenges, especially when using conventional computing machines.

The HH model incorporates the membrane potential and includes the concentration of various chemicals, inside the neuron, in the calculations to represent its behavior. The computational complexity of conductance-based models is orders-of-magnitude higher than that of IaF models, posing a significant challenge for their efficient simulation. For HH models, GPU implementations have been shown to be less efficient compared to reconfigurable hardware solutions [22, 23], even though providing notable speedups [24]. In [12], a simplified version of the HH model is used in an FPGA-based simulator that is able to simulate 400 physiologically realistic neurons on a Virtex-4 FPGA device. Until recently, most HH models accelerated in reconfigurable platforms, due to their efficiency, employed fixed-point arithmetic. However, limited accuracy of fixed-point representation results in a faulty representation of neural spike location, altering the functional behavior of the neuron [25]. The system in [11] simulates a biophysically accurate representation of the neuron using floating-point arithmetic, and the HH model. The cost of the biophysical accuracy is a low network size; the largest system proposed contains four neurons. In our previous work [14], we could simulate 48 *extended* HH neuron model cells (highly biophysically accurate model [26]) with floating point arithmetic on a Virtex 7 FPGA platform. In [23], a similar system is refined to include up to 96 neurons.

In this book, we propose an efficient multi-chip dataflow architecture for the *extended* HH neuron cell and subsequent interconnected network [27], which exploits data locality and minimizes network communications over one or multiple FPGA devices. The proposed system provides several key aspects compared to existing approaches:

- Close to linear growth in the communication cost: with proposed data localization scheme and the resulting linear growth in communication cost, 31$\times$ to over 200$\times$ more neurons could be simulated in comparison to the state-of-the-art designs, which are limited by the exponential growth in the communication cost.
- The extendibility of the system over multiple chips to build more accurate systems: the system maintains linear growth up to eight FPGA devices.
- The use of double floating-point arithmetic for the most biologically accurate cell behavior simulation and an easy implementation of various neuron network topologies, cell communication schemes, as well as models and kinds of cells.
- A high run-time configurability, which reduces the need for resynthesizing the system. Additionally, adaption of routing tables and changes to the calculation parameters are also possible. In this way, the system reduces the time required for experiments with biophysically accurate neurons.

- A powerful simulator designed for high-precision spiking neuron network simulations, but flexible enough to be used for smaller neural networks. The simulator features configurable on- and off-chip communication latencies as well as neuron calculation latencies. All parts of the system are generated automatically based on the neuron interconnection scheme in use. The simulator allows exploration of different system configurations, e.g., the interconnection scheme between the neurons and the intracellular concentration of different chemical compounds (ions), which affect the mode of initiation and propagation of action potentials.

1.2 The Inferior Olivary Nucleus Cell

1.2.1 Abstract Model Description

The neuron cells considered in this book are located in the inferior-olivary nucleus (ION). The ION is an especially well choreographed part of the brain [28, 29]. The *extended* (by gap junctions) HH model based on experimental findings in [29] (Figure 1.1) implements a neuron with three distinct compartments: the dendrite, the soma, and the axon. The gap junctions are part of the dendritic compartment; consequently, the dendritic compartment receives the extra input coming from the inter-neuron connection. The gap junctions

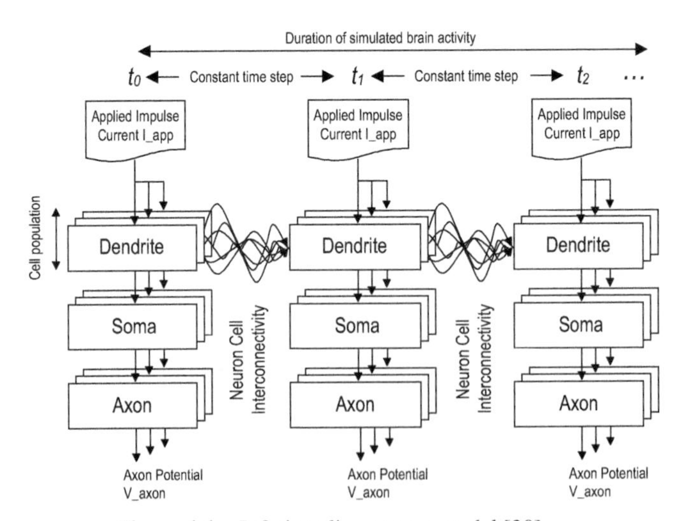

Figure 1.1 Inferior olive neuron model [30].

(which differ from typical synapses in that they are purely electrical) are associated with important aspects of cell behavior as they are not just simple connections, but they involve significant and intricate electrical processes, which is reflected in their modeling details.

Every compartment includes biophysical attributes, i.e., state parameters denoting its electrochemical state, and computation is performed in all three compartments (and within each gap junction's connection itself). For the calculation of a single parameter, one exponential function and several multiplications and divisions need to be carried out. Additionally, for realistic signal representation, the use of floating-point (FP) arithmetic is essential [25]. The total number of FP operations required for simulating a single step of a single neuron cell (including a single gap junction) is 871 (Table 1.1). In an n-cell network (n_c), if each neuron maintains a constant number of connections λ to neighboring cells, the complexity of the overall gap junction computation cost increase as $O_{gj}(n_c \times n_c \times \xi)$, where ξ is the connectivity density [30]. The worst-case interconnectivity scenario occurs when $\xi = 1$, i.e., all-to-all neuron connection, resulting in $O_{gj}({n_c}^2)$ complexity. All remaining, non-gap junction computation increases linearly $O_{cell}(n_c)$ since the rest of the application is of purely dataflow nature [30]. The neuron model defines effectively a transient simulator through computing discrete output axon values in time steps that, when integrated in time, recreate the output response of the axon. The three compartments and gap junctions are evaluated/updated concurrently at each simulation step. The model is

Table 1.1 Neuron requirements per simulation step

Computation	FP Operations per Neuron
Gap junction	12 per connection
Cell compartment	859
I/O and Storage	**FP Variables per Neuron**
Neuron states	19
Evoked input	1
Connectivity vector	1 per connection
Neuron conductances	20
Axon output	1 (Axon voltage)
Compartmental Task	**% of FP Ops for 96 Cells**
Soma	13
Dendrite	10
Axon	8
Gap junction	69

calibrated with a simulation time step of 50 μs, which also defines the real-time behavior of the whole network. Simulations steps are identical to each other in terms of operations performed.

1.2.2 The ION Cell Design Configuration

Operationally, the neuron network needs to compute and communicate simulated ION responses to their neighbors and the axon. We run both operation concurrently and devise separate hardware architectures for computation (based on the multi-compartmental *extended* HH) and communication. We refer to a neuron computation unit as a physical cell (*PhC*)[1]. Within a *PhC*, the topology-dependent (i.e., incorporating the neighbors coupling) dendrite calculation (η_{dend}) and the topology-independent *Axon+Soma* (η_{a+s}) calculation run in parallel (Figure 1.2) [31]. The total amount of cycles each *PhC* requires (η_{PhC}) is

$$\eta_{PhC} = \max(\eta_{dend}, \eta_{a+s}) \tag{1.1}$$

The *Axon+Soma* computational unit computes the axon and soma state and updates a set of cell parameters, based on the current cell compartment states

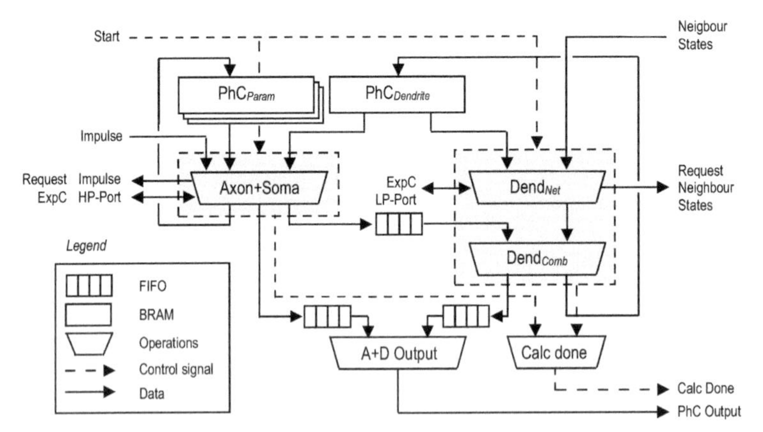

Figure 1.2 Dataflow of a *PhC*. The dashed box on the left is the Axon/Soma calculation. The one on the right is the dendrite calculation. Cell states are stored in the BRAM (memory). All input and output data signals are connected to FIFO buffers (not shown) [31].

[1]The computation units are called physical cells (*PhCs*) to recall that they are physically implemented in hardware, and that the outputs of their computations mimic the actual inferior-olivary nucleus (ION)-cell behavior. See [29] for additional information.

and cell parameters. Internally, the dendrite calculation is dependent on the result of the *Axon+Soma* calculation to calculate the new dendrite state. Externally, both calculations use the same exponent co-processor (*ExpC*). The exponent operations, compared to standard operations require relatively more resources and cycles to complete.

To reduce the required amount of resources without adverse effect on the calculation latency, we utilize a single exponent instance over multiple neuron calculations [31] in a Kahn process network [32]. As the *Axon+Soma* calculation has a longer critical path (and is topology-independent), it is scheduled with a higher priority. Each calculation within a cluster is synchronized, resulting in *Axon* potentials being calculated at predictable times. The dendrite calculation is, however, not synchronized, giving it more flexibility over the exponent co-processor. By keeping the critical path within the dendrite calculation to a minimum, and by allowing it to start processing new network neighbors before the current exponent is known, each simulated cell can quickly be scaled up to allow more connections within the neuron network. The exponent co-processor is, thus, constantly being given new values to calculate, with high-priority tasks arriving after a deterministic amount of cycles.

The Axon+Soma Calculation Unit Configuration: Exponent operands within the *Axon+Soma* calculation unit consist of an addition and multiplication carried out by two constants (χ, β) on a single-cell state (compartmental potential):

$$e^{\phi}, \phi = (State + \chi) \times \beta \tag{1.2}$$

The controller for the application-specific co-processor (ASC) ensures that the correct state potential and operand constants (χ, β) are stored in the registers at the correct time. Multiple *PhC*s are scheduled around a single *ExpC* to reduce the required resources. Consequently, the ASC is adjusted to receive multiple cell states from multiple sources and to send a source address with the exponent operand as (addressed) output. The calculation can be subdivided into two segments: fetch and schedule. Before any calculations can be carried out, the hardware necessitates to *fetch* the required cell states and parameters from the local memory, while also sending out a request to the memory controller to receive a (non-zero) impulse value. With the fetched values, the *Axon+Soma* calculation unit can partly offload its calculations to the ASC, while starting calculations that are independent of the results from the exponent calculations. The η_{a+s} is determined by the number of *PhC*s that share a co-processor, and how the axon and soma calculations are scheduled

$$\eta_{a+s} = \eta_{a+s(base)} + \gamma \times (\nu - 1) - \theta(\nu) \tag{1.3}$$

where $\eta_{a+s(base)}$ is the base latency, i.e., one *ExpC* is connected to only one *PhC*, γ is the number of exponent calculations required by the *Axon+Soma* calculation unit, ν is the maximum number of *PhC*s sharing an *ExpC*, and θ is the overlapping factor.

The Dendrite Calculation Unit Configuration: The *dendrite* calculation unit computes the new dendrite compartmental state based on the current dendrite state, the neighboring dendrite states, and finally an intermittent response generated by the *Axon+Soma* calculation unit. Dend_{Net} computes the coupling effect of neighboring cells in the neuron network; its input is determined by the current dendrite state that is *fetched* from the BRAM memory, and the neighboring dendrite states requested from the memory controller. Dend_{Net} is scheduled around the exponent operation and is split into two parts: computations that are dependent of the exponent result and those that are not. Dend_{Comb} combines the intermittent response received from the *Axon+Soma* calculation unit and results from the Dend_{Net} to generate a new dendrite state potential. The resulting dendrite state from this operation is then locally updated and communicated to the memory controller. The dendrite compartmental computation latency η_{dend}[2] can be written as a function of:

$$\begin{aligned} \eta_{dend} &= \max(\eta_{din}, \eta_{a+s}) + \tau \\ \eta_{din} &= \delta(\varphi) + \max(\eta_{block}, \alpha \times N_D) + \omega \times N_D \end{aligned} \tag{1.4}$$

where α is the amount of cycles that takes place before each (low priority) exponent calculation, N_D is the number of dendrites, δ is the start-up delay partly dependent on the amount of dendrite calculations that share a memory core, and φ is the grouping factor. A blocking time η_{block} is involved if the exponent calculation is being blocked by another task after all α calculations are performed, and ω is the number of cycles after the result of the exponent calculation is known.

The Exponent Core: We schedule the exponent operands through a *read scheduler*, i.e., the scheduler that feeds exponent operands onto a single channel (vector) with an (additional) address[3]. The vector is fragmented over

[2]Within the dendrite calculation, Dend_{Comb} combines the two results after τ cycles.

[3]The high priority is already addressed by the ASC. Low-priority inputs are only passed as operands, and are given an address based on which FIFO they are read from.

the architecture that calculates the exponent, and a shift register that keeps track of where the current calculation (address) is in the pipeline. When the exponent is calculated, a valid address is presented at the end of the shift register, signaling a write back to a specific (addressed) output FIFO.

1.2.3 The ION Cell Cluster Controller

The neuron cells are connected with decreasing probability the further they are apart [1]. The individual computation units, i.e., physical cells that are in a close proximity to each other are placed within a confinement of a (neighbor) cluster. The amount of clusters κ implemented in the FPGA is based on the critical resources and is determined as $\kappa = PhC_{tot}/\varphi$. The cluster controller relates new values to the calculation architecture when requested and stores and routes their responses. Each cluster controller is designed around several parallel running hardware architectures, which are synchronized by FIFOs. In Figure 1.3, an example of a cluster controller is shown with two connected *PhCs*. The *init* controller receives a coded set of initialization parameters (e.g., the cluster identification number, a local routing table, initial parameters for the local *PhCs*, and the dendrite states of all cells) through the initialization channel. The write controller communicates with the *PhCs* by request.

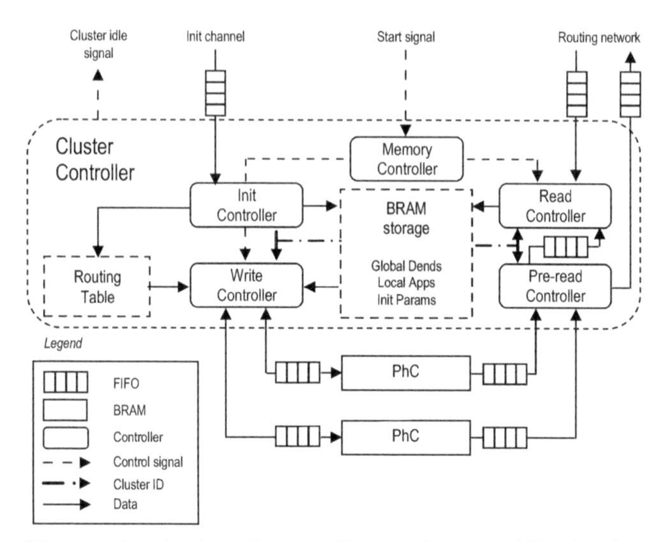

Figure 1.3 Diagram showing how the controllers are housed within the cluster controller; the cluster done logic is excluded from view. Dataflow of a *PhC*. The dashed box on the left is the Axon/Soma calculation.

If the *PhCs* are not initialized and a request arrives, then the write controller transmits the initialization parameters in a pre-defined order starting with the dendrite state. During the simulation, if a dendrite request is received, the write controller looks up (i.e., with the help of the local routing table) which dendrite states of the neighboring cells it should send. Consequently, the write controller sequentially sends each dendrite state addressed on the columns of that row. If an applied current request is received, the local address is used to get the applied current from the BRAM storage. The neighboring cell addresses are placed column-wise in the routing table, while each row represents the location of the local cells in the IO topology. After a cell response is generated, the pre-read controller determines the global address of the cell within the neural network.

If the cell response is an axon value, then the signal is sent to the routing network; if it is a dendrite response, then the new value is duplicated and sent to both the routing network and the read controller through an internal FIFO for storage. The read controller authorizes storage of the applied ION currents and dendrite responses in the BRAM, i.e., the read controller determines which ionic currents will be admitted by calculating the relevant global address range based on the cluster identification number and the amount of cells that are connected to the cluster controller. The dendrite states and applied currents are stored in the BRAM in two parts: current-state and next-state memory. At the start of each simulation round, a start signal is received by memory controller and the bit is issued, which indicates that the next-state memory block is now current state and vice versa. This prevents memory being overwritten by the read controller before it can be sent to the *PhC*. The cluster controller falls into an idle state when the connected *PhCs* have finished calculations and all newly generated results have been stored and/or sent to the routing network.

1.3 Multi-Chip Dataflow Architecture

Neural connectivity have been previously implemented through shared bus networks [14]; however, bandwidth restrictions limit the scalability of such approaches. Alternatively, local buses between adjacent neurons arranged in one-dimensional [19] or two-dimensional [21] grid network have been proposed for increased routing flexibility. However, one of the most notable features of the real brain networks is their high degree of clustering, with nodes (neurons) connecting preferentially to others in their local neighborhood [33]. The high density of local connections in brain networks may have several functional and evolutionary benefits, such as enhanced communication speeds and minimal wiring and metabolic costs. To scale communication linearly with neuron count, we emulate the cortex's hierarchically branching wiring patterns. In the clusters, configurable routing tables define how *PhC*s are arranged within the neuron network. By attaching each cluster to a binary tree network, responses between *PhC*s are shared (Figure 1.4) [27]. Furthermore, through the top node of the tree network, a current impulse can be applied to all *PhC*s, and all output results of the neuron network are streamed.

Localize Communication Between Clusters: The data from other cells is read seldom in the *PhC* [1, 14]. Consequently, a single cell does not require memory access in each clock cycle, allowing for a shared memory design with time-shared instead of parallel memory access. The main advantage is that the common case of close communication is still optimal. The number of *PhC* around one shared memory is limited by placement and wire-length constraints of the FPGA technology in use.

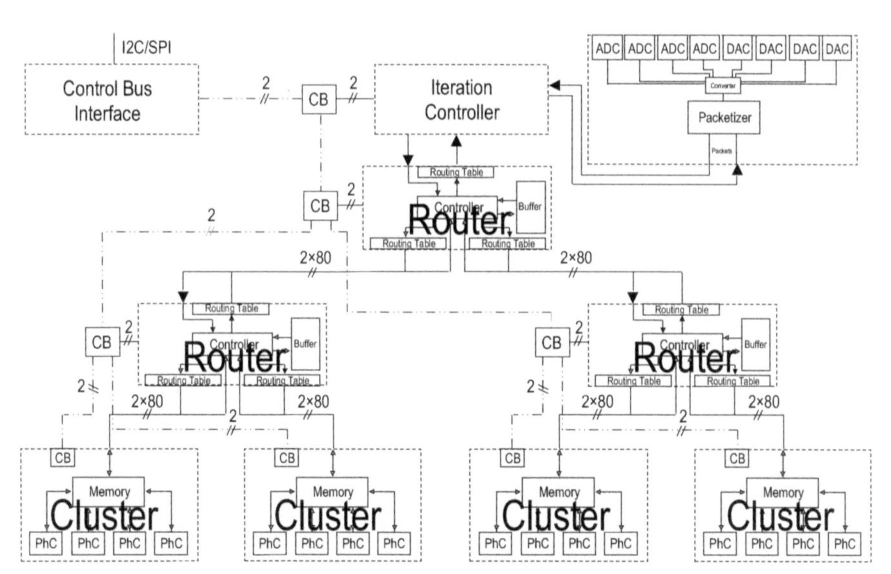

Figure 1.4 System overview. The computing elements (the *PhCs*) are grouped inside a cluster to make communication between neighboring cells fast. These clusters are connected in a tree topology NoC. The router fan-out in this case is 2, which can be changed according to the requirements of the implementation. The same holds true for the number of *PhCs* in any cluster [27].

Connecting Clusters: Routers: In the proposed architecture, each router has two to *n* children, and each child can be either a cluster or another router. The clusters transmit only two types of data, i.e., dendritic and axon hillock potentials. While cell dendrite potentials are shared among all IONs, axon hillock potentials are only given as an output. Consequently, the router is designed with the following rules: i) in a balanced tree network, each router is connected to one bi-directional upstream and two bi-directional downstream channels; ii) new dendrite potential values can arrive through any channel and are passed along the other two channels; and iii) new axon hillock potential values only arrive through one of the two downstream channels and are then transmitted to the upstream channel.

The data produced by each cell in the network and the cell identification number and are combined in a packet. Based on a static routing table (which reflects the way the cells communicate), each router decides in which direction, i.e., to which cells, the packet has to be forwarded to. Within the proposed design, each router (Figure 1.5) is connected to three channels and is implemented around a single core (Router Logic) together with a (FIFO) buffer. The channels consist of an input and output FIFO, forming a bi-directional channel. The router logic reads every channel in a round robin-type fashion. If a new packet is present in one of the channels it is read (and based on the rule set), then the packet is transmitted to one or two of the other channels. However, due to hardware limitations, a channel might fill up before it is emptied (read). Since no packet is allowed to be dropped, packets that cannot be forwarded right away, i.e., when the receiving buffer

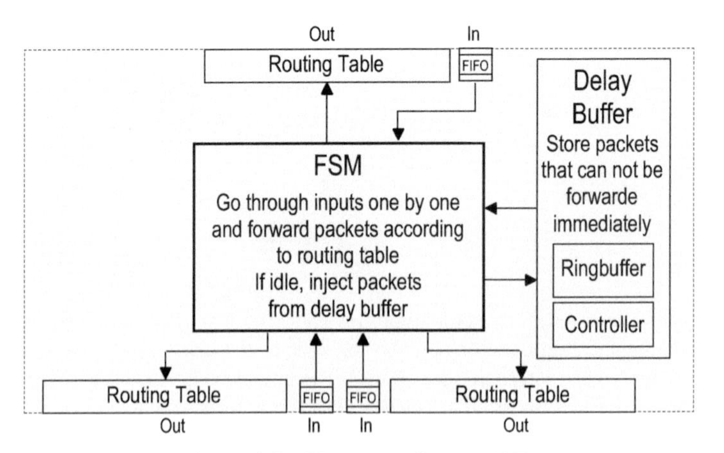

Figure 1.5 The router diagram [27].

is full, are stored for delayed delivery. The width of this delayed buffer is b_p+[$\log_2(n_o)$] bit, where b_p is the amount of bits for a packet and n_o is the amount of outputs of the router. By designing the router around a small finite state machine (FSM), each symbol can be passed to one or two channels every two clock cycles. To avoid cases in which the router continuously tries to deliver delayed packets to full routers, new packets always have precedence over the delayed ones. Since packet forwarding is not aware of the complete network connectivity, the components are efficient and have limited overhead.

The Control Bus for Run-Time Configuration: Due to the large number of components, the most commonly used bus systems, e.g., Wishbone or Serial Peripheral Interface (SPI), are not applicable. In addition, these buses require a significant number of wires to address every component in the system. Consequently, we designed a custom-made bus, which follows the tree structure of the NoC. By traversing the tree, all components can be addressed, from the routers down to the clusters and to each *PhC*. The importance of throughput is reduced, since the configuration is set when the system is paused. The amount of data transmitted via the bus is low with the largest transmissions being parameter changes of a cell. A bus command comprises two parts: the address and the payload. Correspondingly, the bus first opens up a connection to a specific component using the address and then forwards the payload to the component. Each component in the system, e.g., a router or a cluster, has an attached control bus router that can either forward the bus signal to any of its children or forward the bus signal to the attached component.

Adjustments to the Network to Scale over Multiple FPGAs: Since the communication frequency decreases closer to the root of the network tree, multiple FPGAs can be connected at the highest level without significant impact on performance. Although adding another tree layer promises easy extendibility, the limited connection possibilities of each FPGA and need for an extra FPGA for routing between the FPGAs containing the clusters, however, restrict their use. Consequently, as most communications occur between neighboring FPGAs, the FPGAs are connected in a ring-based topology (Figure 1.6), which is less complex in terms of topology generation and administration of the routing tables.

To synchronize the communication between the clusters, one of the FPGAs contains a controller that handles all the synchronization packets. In large systems, this could impact the time needed to complete the iteration. To prevent this, we use one of the FPGAs as a master. Consequently, the signal does not have to cross multiple stages, the run time is constant for

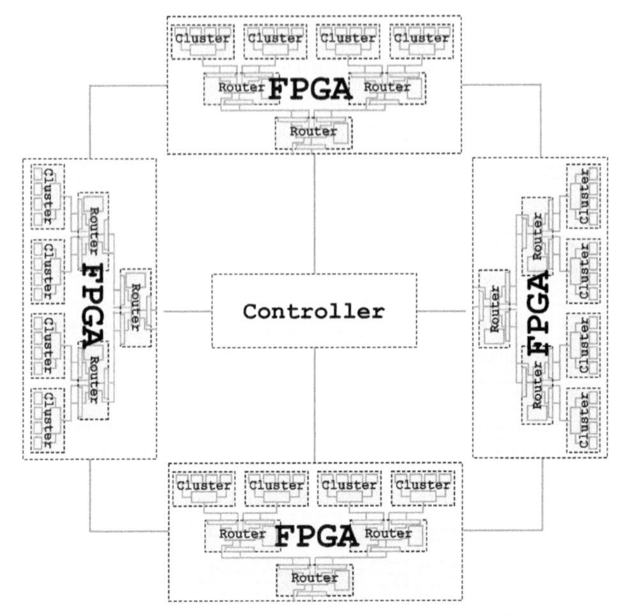

Figure 1.6 Single FPGA implementations are connected using a ring topology network. The FPGA are synchronized via a central controller.

any number of cells, and signal can finish iteration immediately. The master FPGA, in turn, issues the new round signal when adequate.

Experimental Results: The system is automatically generated using a human-readable configuration file, which includes all relevant parameters of the system and can be easily modified allowing exploration of different cell communication schemes, several fan-out values, etc. The control interface includes initialization, setting of I_{app}, direct memory access (DMA) (scatter mode), Ethernet user datagram protocol (UDP) transfer from the FPGA to the PC, and interrupt support. After the design is configured with the desired accuracy (32/64-bit), it is synthesized through the Vivado HLS tool to generate VHDL code and test bench files. The multi-FPGA system experimental setup is illustrated in Figure 1.7, while FPGA resources utilization is shown in Figure 1.8.

A visualization of the ION axon potential V_{axon} run on the FPGA for 1 s brain simulation time (with associated neuron parameters) in a biophysically accurate neuron network consisting of 768 *extended* Hodgkin–Huxley neuron cells is shown in Figure 1.9. Here, an impulse of -1 mA/cm^2 is applied

Figure 1.7 Multi-FPGA system experimental setup.

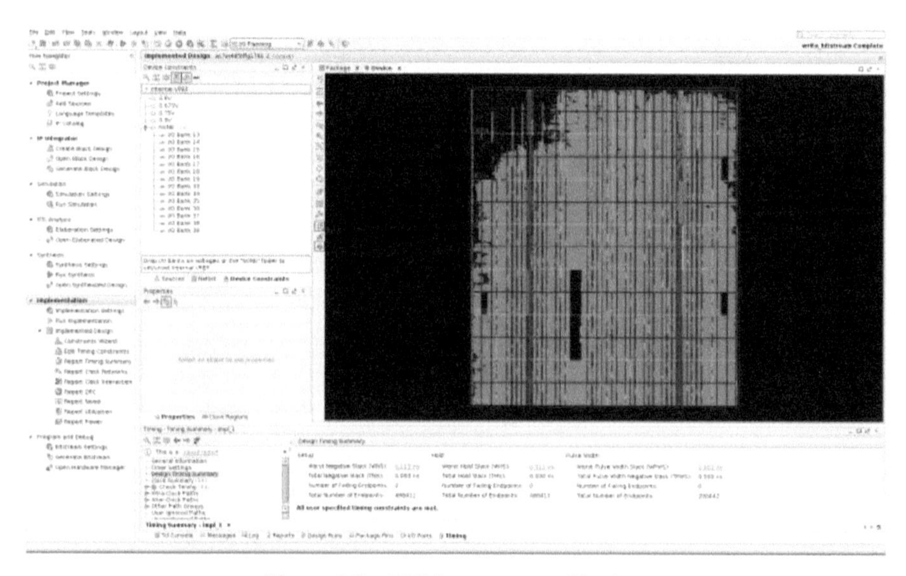

Figure 1.8 FPGA resources utilization.

after 0.19 s for a duration of 100 ms with a resting neural network surface current of 0.5 mA/cm^2. Similar patterns are found with biological test [39]. Simulating identical network settings in SystemC require 59 min of *cpu*-time on an openSUSE 13.1 (x86_64) system with Intel® Xeon® CPUs E5-1620 3.5 GHz processor and 32 GB of memory. Consequently, a FPGA ported design yields >3500× speed-up (performed in real time). The hardware

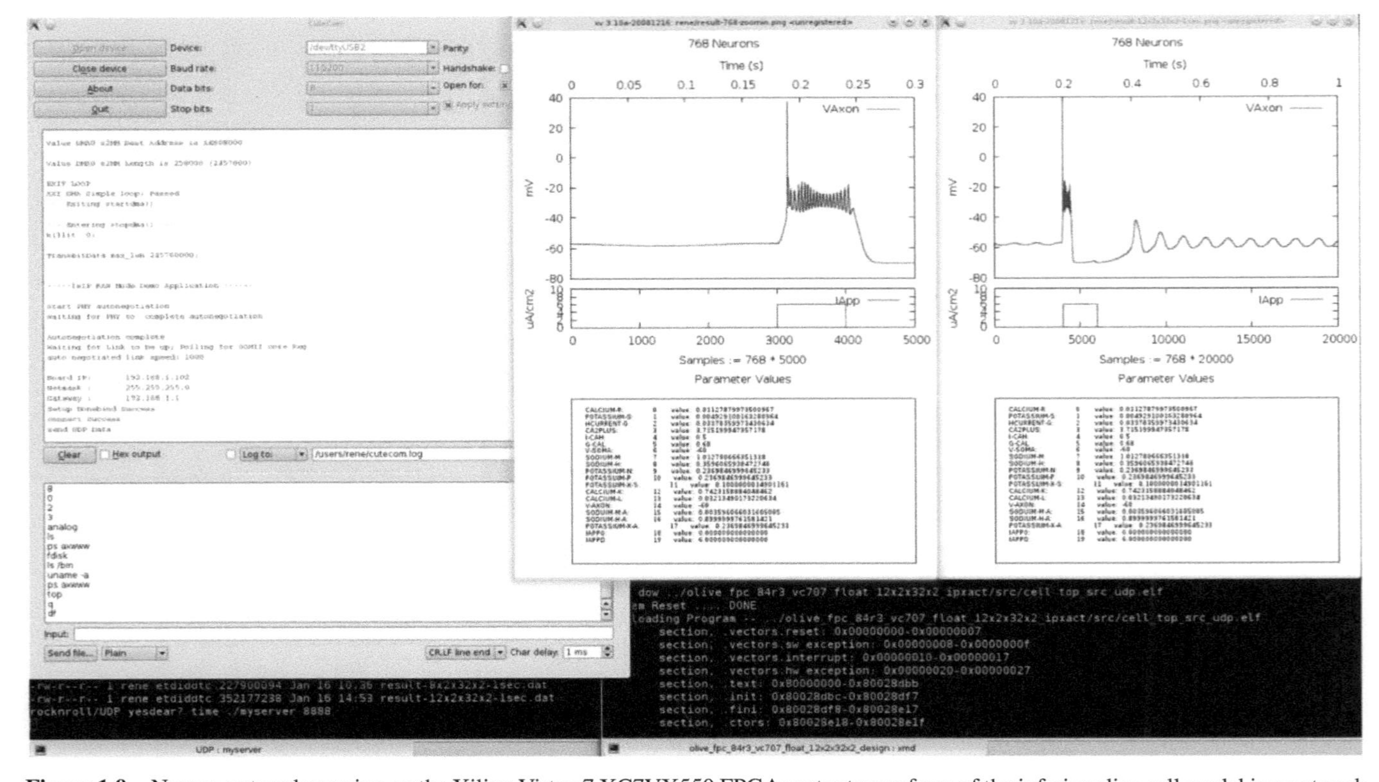

Figure 1.9 Neuron network running on the Xilinx Virtex 7 XC7VX550 FPGA: output waveform of the inferior olive cell model in a network configuration, axon potential vs. brain simulation time.

results are compared to a golden-reference file, containing the expected values of the simulation. Observed error is very low, less than $0.2 \times 10^{-5}\%$ for cell resting state (when most internal cell variables change rapidly), and at cell firing state, for both 32- and 64-bit configurations.

1.4 Organization of the Book

In Chapter 2, a new system architecture is presented, which is able to bridge the gap between biophysical accuracy and large numbers of cells. By localizing communications, the communication cost is reduced from an exponential to a linear growth with the number of simulated neurons. As a result, the proposed system allows the simulation of a large number of the biophysically accurate neuron cells (19,200 cells for neighbor connection mode and over 3000 cells in normal connection mode). An added advantage is that the system can be extended over multiple chips without significant performance penalty. The system is very flexible and allows to tune, during run-time, various parameters, including the presence of connections between neurons, eliminating (or reducing) re-synthesis costs, resulting in much faster experimentation cycles. All parts of the system are generated automatically, based on the neuron connectivity scheme. The proposed simulator incorporates latencies for on- and off-chip communication, as well as calculation latencies. As a result, the resulting highly adaptive and configurable system allows for biophysically accurate simulation of massive amounts of cells.

In Chapter 3, to achieve both accuracy and real-time speed, a complex biophysically meaningful mathematical model has been analyzed and scheduled on a highly pipelined and parallel running architecture design, specified within a SystemC specification. This has contributed to the creation of hybrid neuron network that executes optimally scheduled floating-point operations that, together with open source IP, has resulted in cost-effective solutions, capable of simulating responses faster or on par with their biological counterparts.

In Chapter 4, several efficient models of the spiking neurons with characteristics such as axon conduction delays and spike timing-dependent plasticity in a real-time data-flow learning network have been described. In addition, the trade-offs between the biophysical accuracy and computation complexity are defined for the different models. The experimental results indicate that the proposed real-time data-flow learning network architecture allows the capacity of over 1188 (max. 6300, depending on the model complexity) biophysically accurate neurons in a single FPGA device.

In Chapter 5, an energy-efficient multipath ring network topology for the neuron-to-neuron communication is proposed. The topology is compared in terms of its mathematical properties with other common network topology graphs, after which the traffic distributions across the network are estimated. As a final characterization step, the energy-delay product of the multipath topology is estimated and compared with other low-power architectures. In addition, a simplified binary tree is suggested as a network layer for handling configuration and input/output data that uses a custom channel protocol without the need for routing tables.

The scalable simulation of neuron communication requires a large amount of computing resources. The high throughput of data pose significant challenges on the interconnect network. In Chapter 6, an efficient, hierarchical dataflow architecture for large-scale biophysically accurate multichip implementation of the neural network is proposed. The network is characterized in terms of the topology, routing, and flow control. To find the efficient network structure, analysis of the throughput is performed for the different network with diverse traffic patterns based on the hopcount and bandwidth. The results indicate that the multicast mesh topology offers 33% improvement in comparison to the unicast counterpart.

In Chapter 7 a novel, energy-efficient, neuromorphic, global ECG classifier applicable for an unsupervised learning is proposed, which offers an effective platform to provide personalized prognosis by combining the heterogeneous sources of available information and identifying meaningful patterns in data. The implemented system utilizes adaptive ECG interval extraction for feature extraction, and correlation matrix for unsupervised feature selection. The resulting features are encoded into spikes and offered to neuromorphic liquid state machine spiking neural network for classification. For each clustering, the silhouette coefficients have been calculated, both based on the selected features and the spike times of the pool for each heartbeat. For compatibility with the wearable devices, the classification system requires only one ECG lead. The experimental results indicate that the proposed design can accurately classify seven heart beat types with an overall classification accuracy of 95.5% at the cost of less than 1 μW.

In Chapter 8, a novel, current-based phenomenological synapse model with power-efficient structures, consisting of efficient synaptic learning algorithms and multi-compartment synapses, has been proposed. A vertical insight is given into the design space of spike-based learning rules in regard to design complexity and biological fidelity. Due to various biological conducting mechanisms, the receptors, namely AMPA, NMDA, and GABAa, demonstrate different kinetics in response to stimulus. The designed circuit offers distinctive features of receptors as well as the joint synaptic function. A better computation ability is demonstrated through a cross-correlation detection experiment with a recurrent network of synapse clusters. The analog multi-compartment synapse structure is able to detect and amplify the temporal synchrony embedded in the synaptic noise. The maximum amplification level is two times larger than that of single-receptor configurations. The design implemented in UMC65nm technology consumes 1.92 pJ, 3.36 pJ, 1.11 pJ, and 35.22 pJ per spike event, for AMPA, NMDA, and GABAa receptors, and the advanced learning circuit, respectively.

In Chapter 9, the main conclusions are summarized and recommendations for future research are presented.

References

[1] W. Gerstner, W.M. Kistler, Spiking neuron models: single neurons, populations, plasticity, *Cambridge University Press*, 2002.

[2] E.M. Izhikevich, "Which model to use for cortical spiking neurons?", *IEEE Transactions on Neural Networks*, vol. 15, no. 5, pp. 1063–1070, 2004.

[3] W. Maass, "Noisy spiking neurons with temporal coding have more computational power than sigmoidal neurons", in M. Mozer et al. (ed.), *Neural Information Processing Systems*, MIT press, pp. 211–217, 1997.

[4] W. McColloch, W. Pitts, "A logical calculus of the ideas immanent in nervous activity," *Bulletin of Mathematical Biophysics*, vol. 5, pp. 115–133, 1943.

[5] K. Cheung, S.R. Schultz, W. Luk, "A large-scale spiking neural network accelerator for FPGA systems", *International Conference on Artificial Neural Networks and Machine Learning*, pp. 113–120, 2012.

[6] D. Pani *et al.*, "An FPGA platform for real-time simulation of spiking neuronal networks", *Frontiers in Neuroscience*, vol. 11, pp. 1–13, 2017.

[7] H. Shayani, P.J. Bentley, A.M. Tyrrell. "Hardware implementation of a bio-plausible neuron model for evolution and growth of spiking neural networks on FPGA", *NASA/ESA Conference on Adaptive Hardware and Syst.*, pp. 236–243, 2008.

[8] G. Smith, C. Cox, S. Sherman, J. Rinzel, "Fourier analysis of sinusoidally driven thalamocortical relay neurons and a minimal integrate-and-fire-or-burst Model", *Neurophysiology*, vol. 83, pp. 588–610, 2000.

[9] G.B. Ermentrout, "Type I membranes, phase resetting curves, and synchrony", *Neural Computation*, vol. 83, pp. 979–1001, 1996.
[10] A.L. Hodgkin, A.F.Huxley, "A quantitative description of membrane current and its application to conduction and excitation in nerve", *Journal of Physiology*, vol. 117, no. 4, pp. 500–544, 1952.
[11] Y. Zhang *et al.*, "Biophysically accurate floating point neuroprocessors for reconfigurable logic", *IEEE Transactions on Computers*, vol. 62, no. 3, pp. 599–608, 2013.
[12] S.Y. Bonabi *et al.*, "FPGA implementation of a biological neural network based on the Hodgkin-Huxley neuron model", *Frontiers in Neuroscience*, vol. 8, pp. 1–12, 2014.
[13] M. Beuler, *et al.*, "Real-time simulations of synchronization in a conductance-based neuronal network with a digital FPGA hardware-core", *International Conference on Artificial Neural Networks and Machine Learning*, pp. 97–104, 2012.
[14] M. van Eijk, C. Galuzzi, A. Zjajo, G. Smaragdos, C. Strydis, R. van Leuken, "ESL design of customizable real-time neuron networks", *IEEE International Biomedical Circuits and Systems Conference*, pp. 671–674, 2014.
[15] N. Qiao, *et al.*, "A re-configurable on-line learning spiking neuromorphic processor comprising 256 neurons and 128k synapses", *Frontiers in Neuroscience*, vol. 9, pp. 1–17, 2015.
[16] J. Luo, G. Coapes, T. Mak, T. Yamazaki, C. Tin, P. Degenaar, "Real-time simulation of passage-of-time encoding in cerebellum using a scalable FPGA-based system", *IEEE Transactions on Biomedical Circuits and Systems*, vol. 10, no. 3, pp. 742–753, 2016.
[17] B.V. Benjamin, *et al.*, "Neurogrid: a mixed-analog-digital multichip system for large-scale neural simulations", *Proceedings of IEEE*, vol. 102, no. 5, pp. 699–716, 2014.
[18] A. Andreou, K. Boahen, "Synthetic neural circuits using current-domain signal representations", *Journal of Neural Computation*, vol. 1, no. 4, pp. 489–501, 1989.
[19] P.A Merolla, *et al.*, "A million spiking-neuron integrated circuit with a scalable communication network and interface", *Science*, vol. 345, no. 6197, pp. 668–673, 2014.
[20] B.U. Pedroni, *et al.*, "Mapping generative models onto a network of digital spiking neurons", *IEEE Transactions on Biomedical Circuits and Systems*, vol. 10, no. 4, pp. 837–854, 2016.
[21] J. Navaridas, *et al.*, "Understanding the interconnection network of SpiNNaker", *International Conference on Supercomputing*, pp. 286–295, 2009.

[22] G. Smaragdos, *et al.*, "Real-time olivary neuron simulations on dataflow computing machines", *Supercomputing*, J. Kunkel, *et al.*, eds., Lecture Notes in Computer Science, pp. 487–497, Springer International.

[23] G. Smaragdos, *et al.*, "FPGA-based biophysically-meaningful modeling of olivocerebellar neurons", *International Symposium on Field Programmable Gate Arrays*, pp. 89–98, 2014.

[24] H.D. Nguyen, Z. Al-Ars, G. Smaragdos, C. Strydis, "Accelerating complex brain-model simulations on GPU platforms", *IEEE Design, Automation, and Test in Europe Conference*, pp. 974–979, 2015.

[25] Y. Zhang, *et al.*, "A biophysically accurate floating point somatic neuroprocessor", *IEEE International Conference on Field Programmable Logic and Application*, pp. 26–31, 2009.

[26] P. Bazzigaluppi, *et al.*, "Olivary subthreshold oscillations and burst activity revisited", *Frontiers in Neural Circuits*, vol. 6, no. 91, pp. 1–13, 2012.

[27] J. Hofmann, A. Zjajo, C. Galuzzi, R. van Leuken, "Multi-chip dataflow architecture for massive scale biophysically accurate neuron simulation", *International Conference of the IEEE Engineering in Medicine and Biology Society*, pp. 5829–5832, 2016.

[28] C.I. De Zeeuw, *et al.*, "Spatiotemporal firing patterns in the cerebellum", *Nature Review Neuroscience*, vol. 12, no. 6, pp. 327–344, 2011.

[29] J.R. de Gruijl, *et al.*, "Climbing fiber burst size and olivary subthreshold oscillations in a network setting", *PLoS Computational Biology*, vol. 8, no. 12, pp. 1–10, 2012.

[30] G. Smaragdos, *et al.*, "Performance analysis of accelerated biophysically-meaningful neuron simulations," *International Symposium on Performance Analysis of Systems and Software,* pp. 1–11, 2016.

[31] G.J. Christiaanse, A. Zjajo, C. Galuzzi, R. van Leuken, "A real-time hybrid neuron network for highly parallel cognitive systems", *International Conference of the IEEE Engineering in Medicine and Biology Society,* pp. 792–795, 2016.

[32] G. Kahn, "The semantics of a simple language for parallel programming," *Information processing*, J. L. Rosenfeld, ed., Stockholm, Sweden: North Holland, Amsterdam, pp. 471–475, 1974.

[33] C. Mehring, U. Hehl, M. Kubo, M. Diesmann, A. Aertsen, "Activity dynamics and propagation of synchronous spiking in locally connected random networks," *Biological Cybernetics*, vol. 88, no. 5, pp. 395–408, 2003.

2

Multi-Chip Dataflow Architecture for Massive Scale Biophysically Accurate Neuron Simulation

Jaco Hofmann

Delft University of Technology, Delft, The Netherlands

The ability to simulate brain neurons in real time using biophysically meaningful models is a critical prerequisite grasping human brain behaviour. By simulating neurons' behaviour, it is possible, for example, to reduce the need for in vivo experimentation, to improve artificial intelligence and to replace damaged brain parts in patients. A biophysically accurate but complex neuron model, which can be used for such applications, is the Hodgkin–Huxley (HH) model. State-of-the art simulators are capable of simulating, in real time, at most, tens of neurons. The currently most advanced simulator is able to simulate 96 HH neurons in real time; limited by its exponential growth in communication costs. To overcome this, we propose a new system architecture, which enables massive increase in the amount of neurons to be simulated. By localizing communications, the communication cost is reduced from an exponential to a linear growth with the number of simulated neurons. As a result, the proposed system allows the simulation of over 3000–19,200 cells (depending on the connectivity scheme). To further increase the number of simulated neurons, the proposed system is designed in such a way that it is possible to implement it over multiple chips. Experimental results have shown that it is possible to use up to 8 chips and still keep the communication costs linear with the number of simulated neurons. The systems is very flexible and allows to tune, during run time, various parameters, including the presence of connections between neurons, eliminating (or reducing) re-synthesis costs, resulting in much faster experimentation cycles. All parts of the system are generated automatically, based on the neuron connectivity scheme. The proposed simulator incorporates latencies for on- and off-chip communication, as well as calculation latencies. As a result, the resulting highly adaptive and configurable system allows for biophysically accurate simulation of massive amounts of cells.

2.1 Introduction

A usable brain simulator for real-life experiments should be able to simulate large parts of the brain, which contain thousands, or even millions of cells. The highly parallel nature of neuronal networks leads to an insufficient scalability on classical von Neumann machines [1]. One possible solution to this problem is the design of highly parallel dedicated hardware. In this chapter, we present a hardware architecture capable of dealing with massive numbers of cells needed for large and accurate brain simulators.

Nerve cells, also known as neurons, are cells inside the brain, which are used to process and transmit signals. They mainly consist of three parts. Through the *dendrites*, inputs from other neurons are received as electro-chemical stimuli. These stimuli are transferred to the *soma*, where they are processed and stored as a membrane potential. If this potential reaches a certain threshold, a spike is transferred via the *axon* to other neurons. The neurons considered in this work are located in the cerebellum and the inferior olivary nucleus (ION). They are important for the coordination of the body's activities [2]. The ION is an especially well chartographed part of the brain [3]. Neurons are treated as computing cores that have n inputs and m outputs. These computing cores require a certain amount of time to complete their calculation. Furthermore, they contain parameters (representing the concentration levels of different chemicals inside the neuron) that influence their calculations. Additionally, the neurons are connected following different patterns such as all-to-all connection, connection between neighbouring cells, or more sophisticated connection schemes based on probability or movement directions.

The design is implemented using SystemC, which allows for cycle accurate simulation of the system at a high level compared to classical hardware description languages (HDLs). Based on the features of the brain, an optimal approach is implemented, which represents the lower bound for the run time of the system. While such a system is not implementable in hardware, certain features can be extracted to be used in the hardware design. The system design is evaluated by running simulations for different parameters and scenarios. Furthermore, an estimate for the hardware resource usage of the design is given. This chapter is organized as follows: Section 2.2 introduces related neuron cell models and integration of these models in the network and presents an overview of the system design and details of implementation decisions. Section 2.3 presents the system design from a high-level perspective, while Section 2.4 offers the details of the SystemC implementation. Section 2.5 contains simulations for different scenarios, including network and multi-chip performance. The section ends with an estimate of the hardware utilisation to illustrate the feasibility of a hardware implementation. Section 2.6 concludes the chapter and offers several suggestions for the future work.

2.2 System Design Configuration

2.2.1 Requirements

The system should be able to achieve brain-real-time of 50 μs for one calculation round. While the cell calculations themselves are static in execution time for an increasing number of cells as they are executed in parallel, the interconnect used in [4] scales exponentially with increasing number of cells. Consequently, the first requirement for the system is *linear scalability*. The network has to be faster than processing all calculations (even for tens of thousands of cells), thus making calculations the dominant *cpu* time-consuming factor. Modelling learning and other various dynamic processes in the brain require *flexibility* of the system. During run-time, multiple parameters need to be changeable, including inserting and removing connections between cells. Furthermore, parameters of the calculations themselves need to be changeable. These parameters, among others, represent the concentration levels of various chemicals in the cell. Being able to change these concentration levels during run time enables us to use the system to explore effects of pharmaceuticals and other environmental factors.

The properties of the brain model should be closely incorporated in order to achieve a high system performance for the given requirements. The most important property of the system in this case is the connection scheme. Neuron cells that are further apart are connected less likely. The distance in this case depends on the model, e.g., geographical distance inside a 2D or 3D cell layout, specific direction of movement. Another requirement to the overall system is imposed by the size of FPGA. To extend the system across the boundary that an FPGA imposes, the system design should incorporate a method of interconnecting multiple FPGAs to form a larger system.

2.2.2 Zero Communication Time: The Optimal Approach

The implementations proposed in this section are optimal with respect to run time, but are not feasible for hardware implementation due to resource constraints. Subsequently, the optimal system design is used to evaluate the absolute limitations of the system as well as deriving a feasible system from it.

The first approach might be to model the neuron interconnection following the structure of the brain, i.e., each cell has physical connections to all neighbouring cells. Whether a neuron cell is a neighbouring cell depends on the connection scheme in use; the connection scheme can be arbitrarily complex – all cells are connected to all other cells, or only direct neighbours in a 2D grid are connected. Each of these connections needs to transfer the complete data of the cell (64 bit in this case). In addition, the system requires synchronisation lines because of the discrete nature of the system. Flexibility requirement introduces new restriction; as new connections cannot be generated during run time, it is necessary to have every possible connection already implemented. Thus, each cell has to be connected to every other cell in the system!

Considering the discrete nature of the system, another optimal system can be designed using a shared memory. Instead of connecting each cell to each other cell, the communication data are stored in a memory that is accessed by all cells simultaneously. This memory would have to have n read and n write ports for n number of cells. However, such a memory does not exist for the amount of cells needed in the system. Furthermore, the necessary amount and length of wiring makes this approach infeasible. Using IEEE754 doubles with 64 bit as data and an address length of 12 bit (4096 cells addressable), each cell needs $2 \times (64 + 12) = 152$ bits to read and write address and data ports. While the corresponding random access memory (RAM) can be modelled on a FPGA using, for example, time-sharing of the block random access memory (BRAM), it still requires too much hardware resources to be feasible. Nonetheless, the shared-memory optimal system can be used to derive part of a synthesizable system, as presented in the following section.

2.2.3 Localising Communication: How to Speed Up the Common Case

The main problem with the shared memory optimal approach presented in Section 2.2.2 is that the shared memory does not have enough ports to handle all cells in parallel. Looking at the calculations of each cell, as implemented in [4], indicates that data from other cells is read seldom in the physical cell (PhC). Thus, a single cell does not require memory access for each clock cycle, allowing for a shared memory design with time-shared instead of parallel memory access. The main advantage in this approach is that the common case of close communication is still optimal. This claim is only true for a certain number of PhC in a cluster. At some point, each PhC has to wait too long for memory access and the performance decreases. Furthermore, the number of PhC around one shared memory is limited by placement and wire length constraints of the FPGA technology used. The exact number of PhC around one shared memory has to be determined by simulation. These PhCs around a shared memory with corresponding control logic is furthermore called a cluster.

Apart from the common case, there is still the un-common case of communicating with cells outside the own cluster that, while not as important with respect to performance, has to be handled correctly. Multiple ways to design such a system exist. The simplest being a bus connection between clusters [4]. Even though these connections are infrequently used, the scalability of the bus imposes a high cost for extending the system with more clusters. Another approach, becoming increasingly popular, is Network-on-Chip (NoC), which promises higher flexibility and performance than classical interconnect approaches.

The following section will give a short introduction into NoC and outline how such an NoC can be incorporated into the system.

2.2.4 Network-on-Chips

Several methods of connecting components in a system exist. The simplest and most straightforward is connecting two components with wires for data and some control lines. While simple, this type of connection is also expensive when multiple components have to be connected and not efficient as most connections will not be used all the time. A better and only slightly more complex approach is the use of a bus. Such a communication system uses a single connection between all components and a bus master organizes the access to the single communication medium. As all components can communicate with each other, a bus is less expensive than dedicated connections. On the other hand, all communications have to time-share the one communication line, which decreases performance in systems with frequent communications. An NoC approach promises higher performance, as well as higher efficiency combined with low hardware costs. Typically, an NoC [5] consist of routers that are connected by wires. Each component of the system is connected to one such router. When a component wants to send data to another component, it injects a packet into the network by passing the packet to its router. To support variable size packets, a single packet is typically split into flits. These flits contain information about their type to indicate if the flit is the head or tail of a packet, or even both at the same time when the packet only contains one flit. Additional information in a flit depends largely on the NoC implementation. Typical additional information contained in a flit, besides the data, is the size of the data contained in the flit, the route the packet has to take, or the destination depending on the routing algorithm in use. Furthermore, information for prioritization of certain data using, for example, virtual channels, can be included in a flit.

These flits are routed to a destination according to a routing algorithm and the topology in use. The topology for the NoC is chosen according to the parameters of the system. These topologies might be enhanced by using express channels to connect further away routers for increased throughput, as shown in [6]. The routing algorithm in use depends on the characteristics of the network and the topology. A simple routing algorithm for a 2D Mesh is called XY-Routing, which is a deterministic routing algorithm that always results in the same route taken for a flit. First, the packet is routed along the X axis, and then along the Y axis to reach the destination. While simple, this type of routing has the flaw of being agnostic to the traffic in the network. Certain routes might be blocked, while others are completely empty. The complexity of routing algorithms can be increased by, for example, considering multiple shortest routes to a target, and then selecting one of these paths randomly for each flit resulting in a better network load.

Even more complexity and efficiency can be added by using heuristics and information about the network load at any given time. Other types of NoC require different routing approaches like static routing tables or source-based routing. For source-based routing, only the source of a flit is included, and the network decides which components should receive the packet instead of the sender. Such a routing algorithm can be advantageous when one packet has to arrive at multiple receivers as it avoids resending of the same packet to different destinations.

2.2.5 Localise Communication between Clusters

As shown in Section 2.3, the cluster localises communication between cells that are close to each other. Similarly, it is preferable for the NoC to also speed up the local communications. The tree topology is suitable for this type of communication. Most packets between clusters will only incorporate the next cluster. In a tree-based structure, as shown in Figure 2.1, it will rarely happen that a packet has to cross multiple layers in the tree. The most important parameter that has to be determined through simulation is the fan-out of the routers. A larger fan-out results in more cluster any single router has to deal with. On the other hand, a lower fan-out will result in higher hardware utilisation by requiring more layers in the tree. The second important aspect for the NoC is the routing scheme. Each packet only includes its source. The packet is then routed to the destinations based on the source of the packet. Each router has to decide to which of its outputs the packet has to be forwarded to. The router does not need to know where the packet will eventually be sent, thus ensuring that each router is small and efficient.

The routers (n inputs and n outputs) have to fulfil certain requirements. One of these input–output pairs is corresponding to the upstream in the tree and the rest to the downstream targets. The upstream channel is not present in

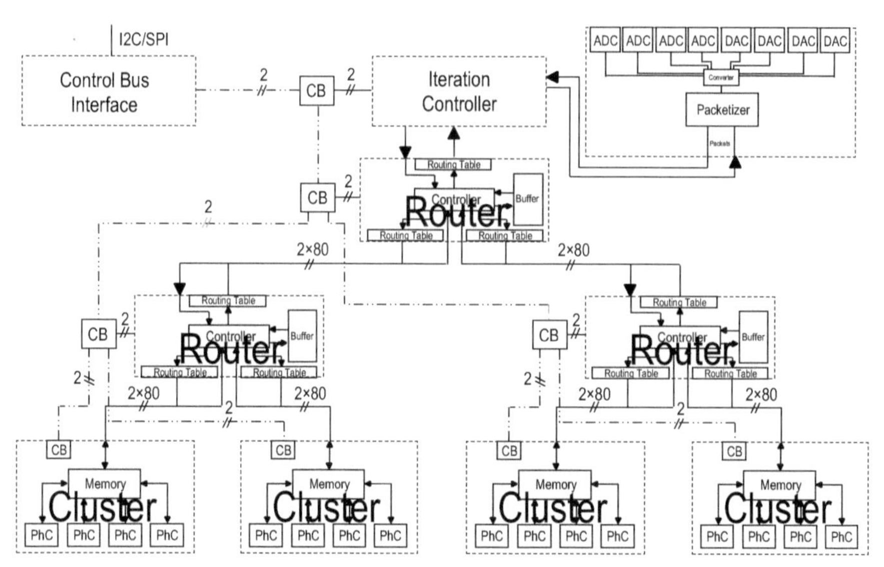

Figure 2.1 System overview. The computing elements (*PhCs*) are grouped inside a cluster to make communication between neighbouring cells fast. These clusters are connected in a tree topology NoC. The router fan-out in this case is 2, and can be changed according to the requirements of the implementation. The same holds true for the number of *PhCs* in any cluster.

the root router of the tree. From each input, a packet might go to each of the outputs, but it cannot be looped back. For each possible source of a packet, the router may decide where to route the packet based on static routing tables. Each packet may be routed to all outputs or to no output at all. In addition, each router needs to be able to deal with congestion. The packet might need to be routed to a router that signalizes it is full. In such a case, the packet has to be sent at a later point, when the receiving router is able to process the packet. A classical approach would be to drop the packet and inform the source of the packet to resend it at a later time. Considering the properties of the NoC, dropping packets has several drawbacks. The router would have to remember that it already got a packet from a certain source, and only forward the packet to the blocking router, thus requiring more hardware. Furthermore, there needs to be an additional channel to indicate to the cluster that a packet was dropped. A better approach for this scenario is to use memory to store all packets that cannot be forwarded right away and try to forward them later at a suitable time. The size of this "delayed" packet buffer can be determined by simulating an all-to-all cell connection scheme, or more specific cases depending on the use case, calculating the highest fill rate of the buffers. Moreover, different optimisations of the routing are possible like preferring upstream packets over downstream packets. The abstract router design is presented in Figure 2.2.

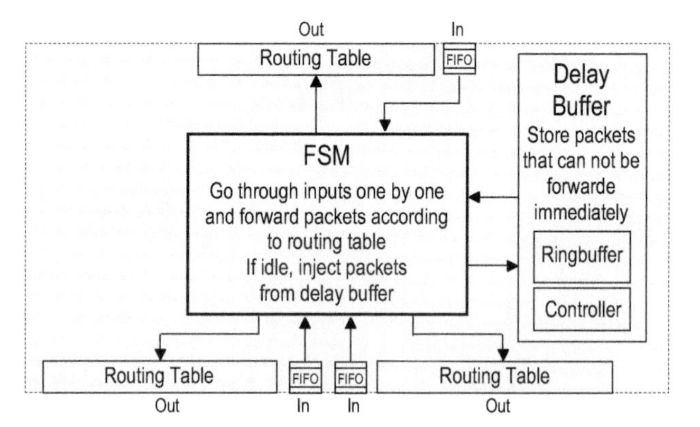

Figure 2.2 The controller queries each input and forwards packets according to the routing tables of the outputs. A packet may be forwarded into any direction, and therefore each output has its own routing table. If a receiving router cannot accept a packet, that packet is stored inside the buffer until it can be forwarded.

2.2.6 Synchronisation between the Clusters

Communications inside one cluster are designed to be much faster than communications between clusters. Hence, a cluster that does not require any communication with a different cluster is finished earlier with an iteration than a cluster that requires such communications. Without any synchronisation scheme, the faster cluster would pull ahead of all the other clusters, which would result in faulty calculations. To solve this issue, different solutions are possible.

The easiest might be a *ready signal* that is set whenever a cluster is ready for the next iteration. A cluster is ready when it sent out all its data, and when all the data it needs from other clusters are received. This signal can be combined using simple or-gates inside the tree, and at the top fed back to the clusters. This approach is very fast, but requires two extra wires in the tree. In addition, there is no control over the "start a new round" signal. The last point can be solved by adding a controller, which fetches the ready signals of the clusters and issues a new round signal at a suitable time. Another possibility is the use of *a special packet that is routed to the root of the tree* where it is counted. After enough done packets have arrived at the root, a start new round packet is issued. This approach has the advantage to utilise the existing infrastructure. A different approach is to *not use synchronisation*, but to incorporate that some clusters are finished earlier than others. The results of the faster clusters can be buffered by the slower clusters and used at the right time. This approach, however, has a higher complexity and does not have any advantages to the system that is designed to run in real time. For pure simulation, it might be interesting when the fast clusters can be reused for a different task after they finished the required amount of iterations. These possibilities are not further evaluated in this work however. Thus, the following three possibilities are discussed here: using a ready signal, routing a special packet to the root of the tree, and not using synchronisation. As the second approach is reusing the existing infrastructure, and the performance is comparable to the first approach, it is used henceforth. The last approach is more complex and does not have any advantage for the current system. Nonetheless, for a task-based system, where the cluster can calculate a different task after they have finished, the last approach might offer particular advantages.

An important aspect for a biologically interesting system is the number of cells. Any single FPGA is limited to a certain number of cells due to hardware constraints. While the system designed in this work promises good scalability to a high number of cells, it is of little use when only one FPGA can be in the system, as the massive number of cells required cannot be accumulated by one FPGA. The following section will characterize the system with respect to multi-FPGA implementations.

2.2.7 Adjustments to the Network to Scale over Multiple FPGAs

To increase the number of simulated neuron cells, integration of multiple chips in one large system is required. However, such a system poses new significant challenges.

The connection between different FPGAs is much less parallel than the connections inside the FPGA. The speed largely depends on the techniques used for transmissions between FPGA and can vary from faster than a transmission between routers inside the FPGA for very sophisticated interconnects to many times slower for simple or low-power interconnects.

It is assumed that the inter-FPGA connection is up to 32 times slower than the communication between two routers inside the FPGA. Furthermore, each FPGA can be connected to only four other FPGAs at the same time. Communication that crosses FPGA boundaries is rare, which follows from the cell connection scheme; consequently, the overall system is well suited for scalable multi-FPGA implementations. The tree topology, on the other hand, is not suited for multi-FPGA systems. While adding another tree layer on top of each FPGA promises easy extendibility, the limited connection possibilities of each FPGA do not allow for it. Extra FPGA need to be used just for routing between the FPGAs containing the cluster. A ring topology promises good performance for this usage scenario. As communication between neighbouring FPGAs in the ring is already rare, communication between FPGAs that are further apart is even less likely. Furthermore, the ring topology generation and administration of the routing tables is less complex compared to a mesh or different topology. Henceforth, a ring topology is used to connect the FPGAs.

Nonetheless, synchronisation between the clusters is still necessary, especially for multi-FPGA implementations. Assuming that one of the FPGAs contains a controller that handles all the synchronization packets, the synchronisation packets need to cross multiple FPGAs to reach that one master. In large systems, this results in a large impact on iteration performance as waiting for the synchronization packet becomes the dominant time consumer. Therefore, a different synchronization scheme has to be used. One of the FPGAs or an extra FPGA is used as master. All FPGAs in the system are connected to this master via two wires. Using this connection, an FPGA can signal a finished iteration immediately. This signal does not have to cross multiple stages, and the run time is constant for any number of cells. The master FPGA, in turn, issues the new round signal when adequate. While this approach requires two extra wires, as well as a dedicated master FPGA, it provides much better scalability and performance compared to the packet-based approach for inter-FPGA implementations. The complete system is shown in Figure 2.3.

2.2.8 Interfacing the Outside World: Inputs and Outputs

While the system uses 64-bit floating point precision, the interfaces to the outside world, i.e., analog-to-digital converter (ADC) and digital-to-analog converter (DAC), typically only have 8–12 bit precision. Thus, the interface has to convert the numbers coming from or going to the converters to the right ranges and input them into the system. The straightforward way of input and output connections are dedicated wires to certain clusters. While this solution is easy to implement and it is fast, flexibility is limited. The clusters that receive inputs and outputs have to be predetermined and cannot be changed during run-time. A better alternative is using the packet system.

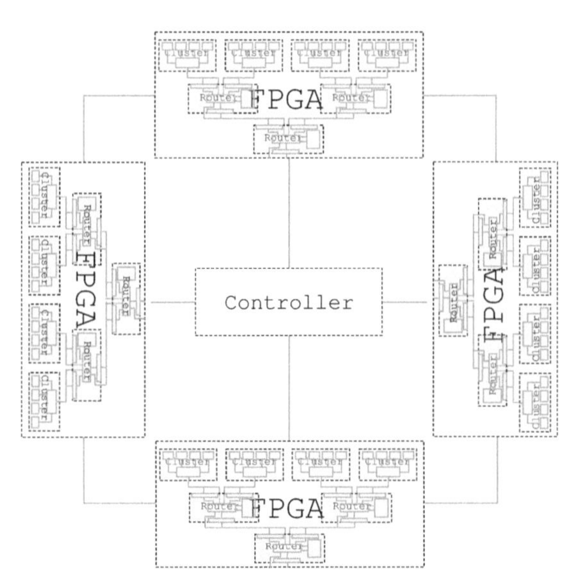

Figure 2.3 Single FPGA implementations connected using a ring topology network. The FPGAs synchronized via a central controller.

The converters can be attached to the top of the tree. For each iteration, the converters send a packet to all cells that require the input and receive a packet from all cells that are meant to output data. This solution is certainly slower than direct connection to the cells, but offers increased flexibility. As most of the communication cost of the packet-based approach can be hidden behind the cell calculations, which results in negligible performance drops compared to the direct wire approach, the packet-based approach is used furthermore.

With the interfacing in place, the mandatory systems are designed. The next section describes the design of a system that enables the user to run-time configure the whole system, including the creation and destruction of intercell connections.

2.2.9 Adding Flexibility: Run-Time Configuration

The system so far is static as no parameters can be changed during run time. For each change in connection scheme or calculation properties, the whole system would have be synthesized with different parameters and reprogrammed. As synthesizing takes time ([4] reports 10 min for a simpler design), experimenting with different connection schemes and calculation parameters becomes tedious. This section describes the design of a configuration bus to allow for run-time change of parameters.

Through the configuration bus, each element in the system needs to be addressable. Thus, each router, cell, cluster, and controller needs an address. Due to the large number of components, most commonly used bus systems, e.g., Wishbone and Serial Peripheral Interface (SPI), are not applicable. Moreover, these buses require a significant wiring to address every component in the system. On the other hand, a well understood bus system is desirable to interface the system with the outer world, as it makes interfacing with the system easier. Inside the system, it is desirable to follow the tree structure of the NoC. Traversing the tree, all components can be addressed, going along the routers, down into the clusters and from there into each PhC. Thus, a widely known bus like SPI or a serial interface like Universal Asynchronous Receiver Transmitter (UART) interfaces the system to the outside world, while inside the system, a custom tree-based bus is used.

The easiest way to design the configuration bus is to use the NoC that is already in place. Although easy, this approach has the disadvantage that the NoC is not designed to address anything but cells. Thus, the address space needs to be increased and the overall performance of the system would decrease. A favourable approach is to use an extra bus that needs to be small.

Throughput is of less interest, as all configuration should happen while the system is paused, and the amount of data transmitted via the bus is low, with the largest transmissions being parameter changes of a cell, which consists of the address and 64 bit for the data.

The run-time configuration bus is the last system designed. The following section summarises all the configuration parameters that are present in the system to illustrate the parameters of the design that have to be explored using the SystemC simulation.

2.2.10 Parameters of the System

The designed system has multiple parameters: these parameters should be controllable without large changes to the system to enable simulation of different scenarios. The first parameter is the amount of cells in the system. The full amount of cells is determined by the amount of cluster and the amount of PhCs in each cluster, as well as the amount of time-shared cells in each PhC. For the NoC, the most important parameter is the fan-out of the routers. Furthermore, it has to be determined how the cells in the system are connected logically. All of these parameters determine the final system structure.

Additionally, there are many more parameters that specify details of the system. To allow for configuration by a human, or automated simulation operator, without the need to recompile the simulation, the system structure should be specified during run time. The human readable file format used is JavaScript Object Notation (JSON), as it has a large user base, and many libraries are available to read and manipulate the format. Specifying the connection between cells by hand is tedious. As all connections between cells are possible, a system of 1k cells has 1k $\times$ 1k possible connections. When considering that the cells are connected using probability patterns, it becomes clear that the task to create these adjacency matrices needs to be automated. The following section describes the script that is used to generate system configurations.

2.2.11 Connectivity and Structure Generation

To ease the generation of configurations for the system, a script is used that automatically generates the JSON files. The script has defaults for all parameters so that only the parameters that are of interest for the current experiment need to be specified. The most important feature is the support for different connection schemes. The cells can be aligned as 2D or 3D mesh. The distance calculation that is used to determine if two cells are connected can be changed and ranges from geometrical distance to more complex distance calculations that favour connections that go into a certain direction. Half-normal distribution is used to determine if a certain connection is present or not according to the distance calculated with the previously mentioned distance calculation method. Furthermore, the connections can be visualized as a Graphviz graph. The script is implemented in Python.

2.3 System Implementation

This section presents the details of how the design devised in the previous section is implemented. The implementation is done using cycle-accurate SystemC to simulate the hardware, while aiming at flexibility to explore the design space and accuracy to allow for meaningful conclusion for a hardware implementation. This section is subdivided into the different parts of the system containing clusters, routers, round controller, the real-time control bus, and the input/output part. Furthermore, the details of the configuration generation script are presented.

2.3.1 Exploiting Locality: Clusters

The cluster implementation is based on the PhC design implemented in [4]. Each PhC calculates multiple neuron cells time-shared as the brain real time of 50 μs is slow compared to the clock speeds of modern hardware. The design is adjusted to better suit the tree-like structure. In the original system, each PhC first completed calculations before proceeding to interchange information with other PhC. For the clusters, both types of operations are done in parallel. This is possible, since not all calculations performed in a PhC require information from other cells. For implementation details of the PhC, please see [4], as this work does not focus on the calculation itself, but on the interconnection between cells given the communication pattern. As shown in Figure 2.4, the PhCs are grouped around a shared memory. The shared memory is in dual-port configuration offering two read and two write ports. Furthermore, the shared memory is double-buffered. All writes are done to one memory, while all reads are done from another memory, whereas after each round, the two memories change their task. This is necessary as the algorithm is iteration based, and the data for the next iteration need to be

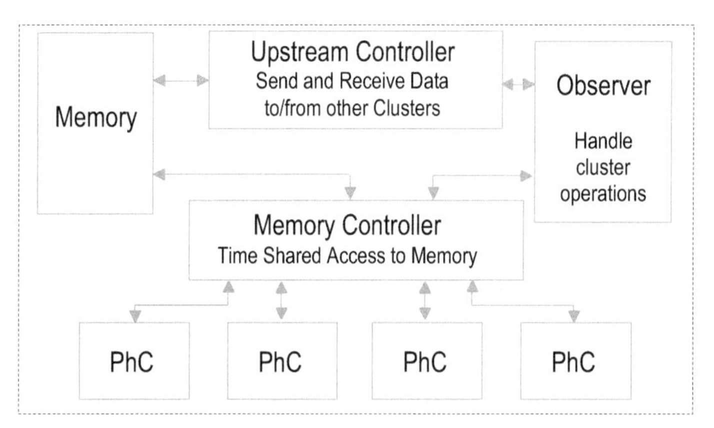

Figure 2.4 The cluster allows for instant communication between any cells that are located inside the cluster. Each of the *PhC* time-shares the calculation of multiple cells. The data of the cells that might be needed by other clusters are forwarded into an NoC.

written to the memory, while the data for the current iteration are still needed for the calculations. The PhCs time-share one port of the dual port memory. Every PhC gains read and write access to the memory using round-robbing. Thus, a PhC has to wait for n clock cycles to gain memory access, where n is the number of PhCs in a cluster.

The cluster is controlled by three controllers. The first controller handles all the book-keeping. This includes monitoring the start signals that are received from the NoC and originate from the round controller. Furthermore, the first controller observes when the PhCs are finished with calculating results. After all data for the next iteration are present in the memory, the book-keeping controller issues a done packet to the round controller. The second controller handles everything related to the NoC. It observes, which packets arrive from the NoC and stores them into the memory using the second memory port. Furthermore, the second controller informs the book-keeping controller that all needed packets arrived. The last controller handles everything that originates from the PhC. It sends all packets upstream into the NoC and controls the round-robbing memory access. Additionally, the third controller stalls the calculation, if the upstream router is not able to receive more packets.

The upstream NoC consists of routers that are responsible to forward the packets originating from the clusters to the correct receiving clusters. The following section describes their inner workings.

2.3.2 Connecting Clusters: Routers

In the proposed architecture, each router has two to *n* children, and each child can be either a cluster or another router. The clusters transmit only two types of data: dendritic and axon hillock potentials. While cell dendrite potentials are shared among all IONs, axon hillock potentials are only given as an output. Consequently, the router is designed with the following rules: (i) in a balanced tree network, each router is connected to one bi-directional upstream and two bi-directional downstream channels; (ii) new dendrite potential values can arrive through any channel and are passed along the other two channels; and (iii) new axon hillock potential values only arrive through one of the two downstream channels and are then transmitted to the upstream channel.

The data produced by each cell in the network and the cell identification number and are combined in a packet. Based on a static routing table (which reflects the way the cells communicate), each router decides in which direction, i.e., to which cells, the packet has to be forwarded to. Within the proposed design, each router (Figure 2.5) is connected to three channels and is implemented around a single core (Router Logic) together with a (FIFO) buffer. The channels consist of an input and output FIFO, forming a bi-directional channel. The router logic reads every channel in a round robin-type fashion. If a new packet is present in one of the channels, it is read (and

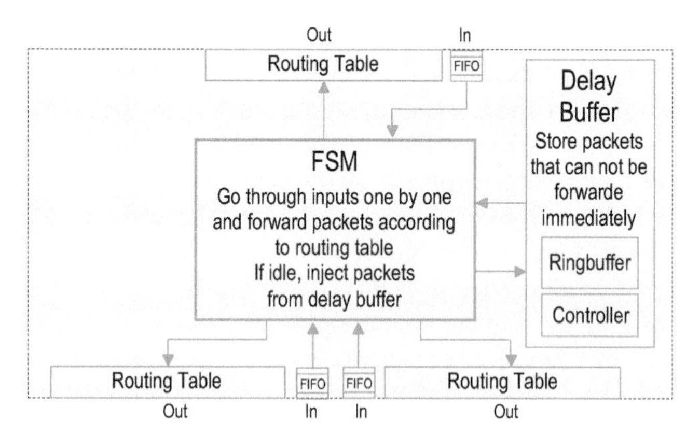

Figure 2.5 The routers are arranged in a tree topology. Each router has *n* children, and except for the root router one upstream router. The router reads input packets from the input first-in, first-out buffer (FIFO). Based on routing tables, it determines where the packets have to be forwarded. If a receiving FIFO is full, the packet is placed in the delayed buffer to be forwarded when the receiving router is free again.

based on the rule set), the packet is transmitted to one or two of the other channels.

However, due to hardware limitations, a channel might fill up before it is emptied (read). Since no packet is allowed to be dropped, packets that cannot be forwarded right away, i.e., when the receiving buffer is full, are stored for delayed delivery. The width of this delayed buffer is $b_p + [\log_2(n_o)]$ bit, where b_p is the amount of bits for a packet and n_o is the amount of outputs of the router. By designing the router around a small finite state machine (FSM), each symbol can be passed to one or two channels every two clock cycles. To avoid cases in which the router continuously try to deliver delayed packets to full routers, new packets always have precedence over the delayed ones. Since packet forwarding is not aware of the complete network connectivity, the components are efficient, and with limited overhead.

2.3.3 Tracking Time: Iteration Controller

The iteration controller is needed to keep all clusters in the system in sync. It is not necessary for the iteration controller to identify that a specific router is done, but only that all clusters have completed calculation. After a cluster finishes a round, which includes completing all its calculations and receiving all data for its next round, the cluster sends a done packet. Any packet that originates from this source is routed upstream to the iteration controller. The iteration controller counts any packet that originates from this source. After the controller received as many packets from this source as there are cluster in the system, it will issue a packet with a source lying one after the done packet. This packet is, in return, routed to all clusters and interpreted as a round start signal.

Next to the iteration controller, at the root of the tree, the system to receive information from and route information to the outside world is located. The next section presents the implementation details about that system.

2.3.4 Inputs and Outputs

The input and output module is used to interface the system with the outside world as shown in Section 2.8. It is assumed that the converters each have an n bit interface and the system can read from, and, respectively, write to these interfaces. The amount of inputs and outputs is not predetermined and depends on the actual system implementation. It is assumed that there are eight inputs and eight outputs. The first step is to convert the n bit data from the inputs to system precision. Each of the inputs receives a unique ID starting from $n_c + 2$ (n_c being the number of cells in the system), thus after the start and stop packets. In each system iteration, these packets are injected into the system. In return, the outputs receive packets from specific cells, which are converted to DAC precision and outputted.

Another interface to the outside world is the control bus, which is used to control the system during run time. The next section describes the implementation of the control bus.

2.3.5 The Control Bus for Run-Time Configuration

The control bus can be used to control the system during run time. The outside world is interfaced using widely known buses like I2C or SPI. Inside the system, a custom tree-based bus is used to utilise the tree-based NoC structure and allow for easy addressing of any component in the system. The bus consists of two wires, one to indicate that the bus is active and one for transmitting the data. Furthermore, the main clock signal is used.

A command consists of two parts: the first part is the address and the second part is the payload. Correspondingly, the bus first opens up a connection to a specific component using the address and then forwards the payload to the component. The payload itself consists of the command that has to be executed and optionally data. Each component in the system, like a router or a cluster, has an attached control bus router. This control bus router can forward the bus signal either to any of its children or to the attached component.

The first step to send any control signal is to send the complete data, including address and payload, over the outside world bus. After receiving the complete bus data, the bits are injected in order onto the control bus router at the top. When a router receives an active bus signal, it will read the first bit on the bus. A high signal indicates that the component corresponding to the router is addressed, whereas a low bit indicates that the router has to forward the signal to any of its children. If the signal is low, the next bits indicate to which child the router should forward the bus data. After the router has been set up for forwarding either to its component or to its child, it will proceed to forward bus data until the bus active signal is low again.

After setting up all routers, the addressed component receives the bus data. This first 8 bits that the component receives are the command that shall be executed. Currently, there are three commands: (i) stop simulation – only valid when round controller is addressed; (ii) replace one bit in routing table – only valid when router is addressed; and (iii) replace any concentration level – only valid when cluster or PhC or cell is addressed. The commands are kept simple on purpose, and more complex commands like creating a new connection between two cells have to be created by combining multiple simple commands. Each command might require additional data. The second command requires the routing table that has to be changed, the entry that has to be changed and the new value for that entry. The third command requires the concentration that has to be changed, and the new value, and the first command does not require additional parameters.

In addition to the two wires used for bus activity and data, there might be a third wire that can be used to stall the bus. The stall signal is used when more data are expected from the outside world, but has not yet arrived. It is preferred, although, to first buffer all the data from the outside world, and then start sending to avoid incomplete commands.

The last part of this section describes how all the components are integrated inside the simulation.

2.3.6 Automatic Structure Generation and Connectivity Generation

To keep the system simulation flexible, a configuration file format is used as described in Sections 2.2.10 and 2.2.11. The configuration file is used to generate the structure of the system, which consists of initialising all components with their respective initialisation data and connecting them. When the simulation starts, the configuration file is read either from a file or from the standard input. The JSON data are then parsed by the Json-CPP library. For any value that is not specified by the configuration file, a default value is used. The parsed data are processed and stored in a configuration structure for further usage. During simulation initialisation, the system structure is built according to the configuration structure. While building the structure, the parameters each component needs are generated. Routing tables are generated in this step considering the position of the router in the tree topology and the adjacency matrix. All components are finally connected using FIFO or signals as necessary.

2.4 Experimental Results

The system is automatically generated using a human-readable configuration file, which contains all the relevant parameters of the system and can be easily modified allowing exploration of different fan-out values, different cell communication schemes, etc. All simulations are performed with cycle-accurate

SystemC, including all calculation and communication latencies, both on- or off-chip. Three different connection schemes are used to simulate the system behaviour; all-to-all connections, normal distributed distance based connections, and neighbour-based connections.

Design Space Exploration: For the router performance, the time when delayed packets get injected into the network is essential: we compared three injection models (Figure 2.6), i.e., packets are injected after each run through all input FIFO, packets are only injected if the FIFO have been empty for a complete round, and packets are injected if there has been no activity on the inputs for 10 rounds.

Injection after each iteration is not feasible and results in very long iteration times. Figure 2.7 highlights the different fan-out in respect to cluster sizes. Small clusters with small routers and bigger clusters with large fan-out (>8) provide the best overall performance. The best overall performance is achieved by 2 *PhC* with a router fan-out of 2 with 4012 cycles. The cluster size choice is illustrated in Figure 2.8. All routers are kept at a fan-out of 2 (i.e., optimal fan-out as shown in Figure 2.7). Due to the fact that all the *PhC*s of a cluster time-share a memory, the clusters with more *PhCs* perform worse, especially at a higher number of cells: the difference for 2048 cells between 2 *PhCs* clusters and 18 *PhC* clusters is 30%.

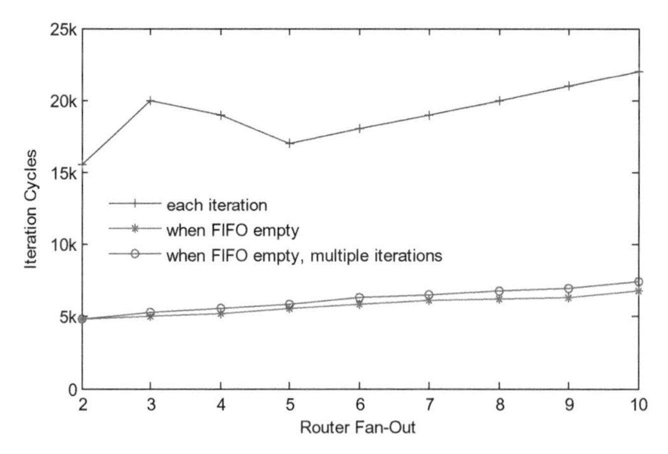

Figure 2.6 Three packet injection schemes. Injecting the packets after each round through all input FIFO results in the slowest iteration times. Injecting only when all FIFO are empty for one round or for multiple rounds achieves comparable performance. Injecting after one round of input inactivity results in the best overall performance.

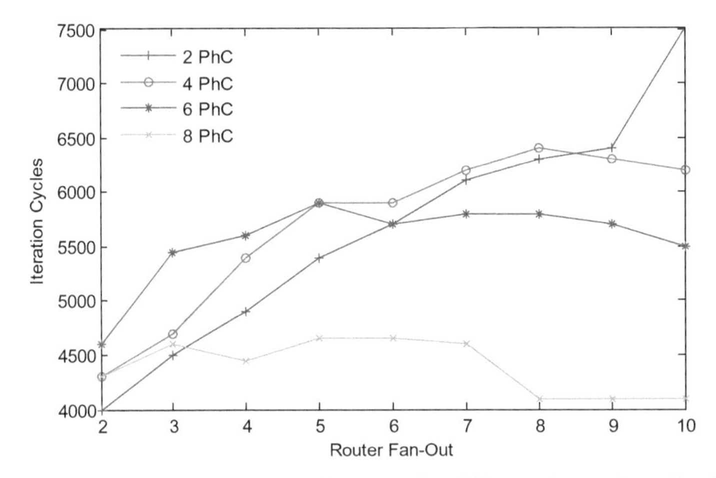

Figure 2.7 Comparison of iteration performance for different cluster sizes. Small routers perform better than large routers – the best performing configuration consists of small clusters and small routers at two *PhCs* per cluster and a fan-out of two.

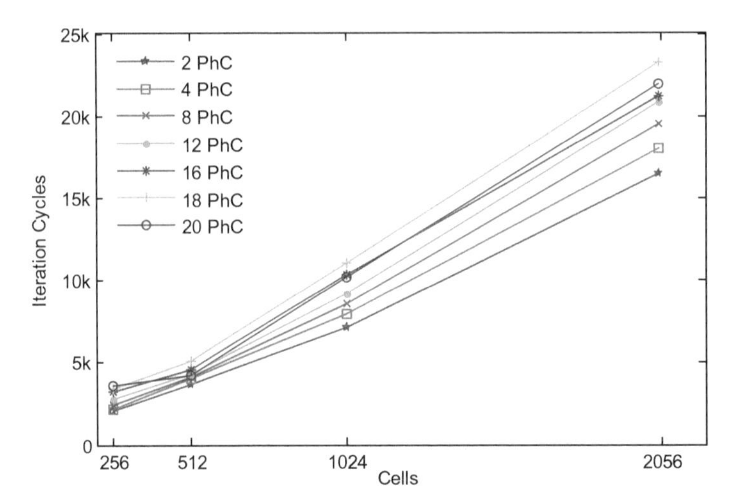

Figure 2.8 Comparison of systems with different cluster sizes; the routers are kept constant at fan-out of two; clusters with two, four, and eight *PhCs* achieve the shortest iteration times, while larger clusters are generally slower.

Single FPGA Performance: A comparison between different cluster sizes for different amounts of cells is illustrated in Figure 2.9. The absolute difference in execution time for 32 to 512 cells is 354 cycles, and 191 cycles in the worst case (2 *PhCs* system). While the proposed system scales *linearly* with the number of cells in the system, the baseline [4] scales *exponentially*.

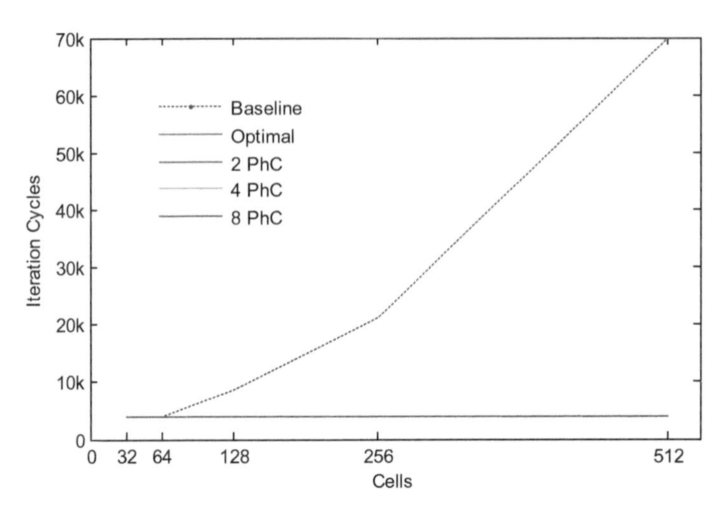

Figure 2.9 Comparison between different cluster sizes for different amounts of cells. The neighbour connection scheme is used and each cell calculation takes 534 cycles. The systems use a router fan-out of two. The baseline design is from [4]; all presented configurations of the new system meet the brain real time at 100 MHz. Furthermore, all of them scale *linearly* with the number of cells. The baseline on the other hand scales *exponentially*.

Considering an average increase in the run time of 120 cycles for the twice the amount of cells, the maximum number of *PhCs* on a single chip can be estimated as $\psi = (c_n - c_o)/120$, where c_n is the number of iteration cycles available for one iteration, i.e., at 100 MHz within the brain real-time boundary,[1] it leads to 5000 cycles, and c_o stands for the amount of cycles currently used, i.e., 4372 cycles at 512 cells. The value received, ψ, denotes the amount of times the number of cells in the system can be doubled: $\psi = 5.23$. Thus, the system supports more than 19,200 cells on one chip for the neighbour connection scheme. For normal connection mode, the system requires 4597 cycles at 512 cells. The increase in iteration time for the twice the amount of cells is 155, leading to $\psi = 2.6$ doublings, and hence, more than 3000 cells on one chip.

Multi-FPGA Performance: Comparisons are performed using both the dedicated wire and the packet-based synchronization methods. In contrast to the packet-based approach, in a dedicated controller method that is connected to each chip, no packet has to cross multiple chip boundaries. For the normal

[1]The maximum number of cell states that can be computed within the model (in the case of the evaluated, high-detail inferior-olive model, the simulation time step is 50 μs [4]).

connection scheme, the dedicated controller performs slightly better than the packet-based controller (Figure 2.10). The dedicated controller system is 3.84 times faster than the packet-based approach at 8 chips with very slow communication (Figure 2.11). For higher connection speeds, on the other hand, the difference in communication time is in the order of 1%.

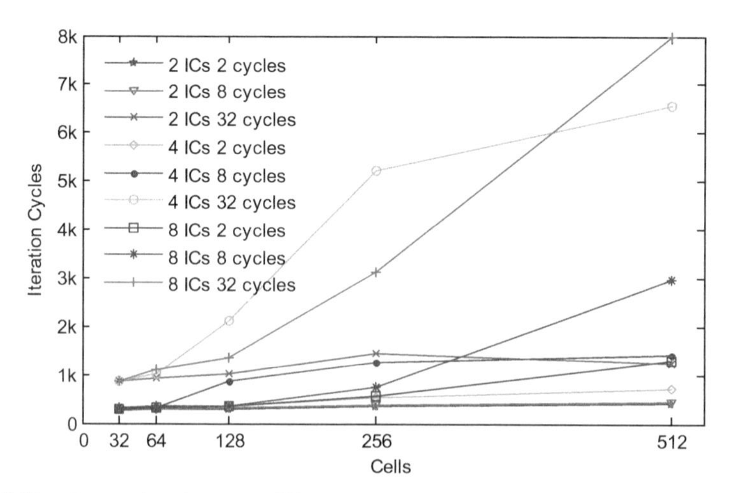

Figure 2.10 Comparison between different system configurations utilizing between one and eight chips; the multichip systems utilize the *packet*-based synchronization method.

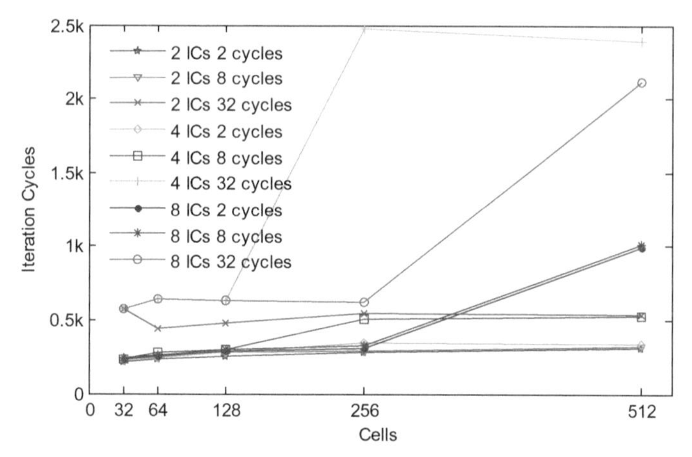

Figure 2.11 Comparison between different system configurations utilizing between one and eight chips; multichip systems utilize the *dedicated wire*-based synchronization method.

2.5 Conclusions

Current neuron simulators, which are precise enough to simulate neurons in a biophysically meaningful way, are limited in the amount of neurons to be placed on the chip, the interconnect between the neurons, run time configurability, and the re-synthesis of the system. In this chapter, we propose a system that is able to bridge the gap between biophysical accuracy and large numbers of cells (19,200 cells for neighbour connection mode and over 3000 cells in normal connection mode). While perfect localization of communication is not possible due to physical constraints, the cells can be grouped around a shared memory in clusters to allow for instantaneous communication. Clusters that are close communicate using only one hop in the network; clusters that are further away communicate less frequently and, consequently, the penalty for taking multiple hops is less severe. The added advantage is that the system can be extended over multiple chips without significant performance penalty. This combination of clusters and a tree topology network-on-chip allows for almost linear scaling of the system. To provide run-time configurability, a tree-based communication bus is used, which enables the user to configure the connectivity between cells and change the parameters of the calculations. As a result, re-synthesizing the whole system just to experiment with a different connectivity between cells is not required. The user has to enter the amount of neurons in the system as well as the desired connectivity scheme. From this information, all required routing tables and topologies are automatically generated, even for multi-chip systems.

References

[1] H.A. Du Nguyen. "GPU-based simulation of brain neuron models". MSc thesis. Delft, The Netherlands: Delft University of Technology, Aug. 2013.

[2] Edward J. Fine, Catalina C. Ionita, and Linda Lohr. "The History of the Development of the Cerebellar Examination". In: *Seminars in Neurology* (2002), pp. 375–384.

[3] Chris I. De Zeeuw et al., "Spatiotemporal firing patterns in the cerebellum". In: (2011), pp. 327–344.

[4] M.F. Van Eijk. "Modeling of Olivocerebellar Neurons using SystemC and High-Level Synthesis". MSc Thesis. Delft, The Netherlands: Delft University of Technology, 2014.

[5] William J. Dally and Brian Towles. “Route Packets, Not Wires: On-chip Interconnection Networks”. In: *Proceedings of the 38th Annual Design Automation Conference*. DAC ’01. Las Vegas, Nevada, USA: ACM, 2001, pp. 684–689.
[6] James Balfour and William J. Dally. “Design Tradeoffs for Tiled CMP On-chip Networks”. In: *Proceedings of the 20th Annual International Conference on Supercomputing*. ICS ’06. Cairns, Queensland, Australia: ACM, 2006, pp. 187–198.

3

A Real-Time Hybrid Neuron Network for Highly Parallel Cognitive Systems

Jan Christiaanse

Delft University of Technology, Delft, The Netherlands

For comprehensive understanding of how neurons communicate with each other, new tools need to be developed that can accurately reproduce and mimic the behaviour of such neurons in real time. By using current complex mathematical models, simulated neurons are able to accurately approximate the behaviour of biological neural tissue. This comes at the price of computing complexity, resulting in responses that lag behind, and thus cannot interface with biological neurons.

The proposed design models an *Inferior Olivary Nucleus* network on an FPGA device, with a maximised amount of simulated neurons for the given FPGA family type. To achieve both accuracy and real-time speed, a complex biophysically meaningful mathematical model has been analysed and scheduled on a highly pipelined and parallel running architecture design, specified within a SystemC specification. This has contributed to the creation of hybrid neuron network that executes optimally scheduled floating-point operations that, together with open source IP, has resulted in cost-effective solutions, capable of simulating responses faster or on par with their biological counterparts.

3.1 Introduction

Neuron networks consist of thousands/millions of neurons, which are highly interconnected via synapses used to transmit signals to individual target cells. To accurately simulate the behaviour of a highly parallel cognitive system such as the ION, a biophysically meaningful model is chosen that closely resembles the biological responses in the human brain [6]. The extended Hodgkin–Huxley model (HH) describes the relation between the electric current to a single neuron membrane and its capacitance. This relation is translated into non-linear differential gap functions [7] that describe the responses of three main parts of a neuron (dendrite, soma and axon). These functions rely a great deal on accurate floating-point operations, and in particular, the exponent operation, as ionic currents in biological neurons follow an exponent trend. Within HH cell, the exponent operation is only used 30 times per neuron calculation. Compared to other more often used operations, the exponent operations requires relatively more resources and cycles to complete. In addition, within the neural network, there exists a high level of connectivity between separate cells. As a consequence of the increased complexity, communication load could increase exponentially, resulting in non-real-time simulation times.

A neural network is implemented on a field-programmable gate array (FPGA) with the help of modern tools and a mix of open- and closed-source IP. Each calculation instance implemented on the FPGA represents a single neuron within the network. By scheduling the calculation elements around the exponential operation, resources are spared, while the required amount of cycles are kept to a minimum. Two approaches are taken to communicate between neurons. First, on a local level, the elements are connected to a bus. Second, these locally connected neurons (cluster) are interfaced in a binary tree network with routers. To test how accurate the model is, a reference model is used to generate neuron responses for a comparative system.

In essence, there are two main drives to simulate an ION on an FPGA. First, the possibility to generate biophysical meaningful responses in real time, and second, to be able to scale up the design and generate as many responses as possible, with the least amount of resources and required power. This came with several challenges, due to the hard real-time deadline and the limited resources available on the chip, and a strong scalable architecture needs to be designed.

Operationally, the SimCs within the design will need to calculate and communicate simulated ION responses to their neighbours and the axon. To simulate the real-time behaviour of a complete neuron network, all implemented SimCs must have carried out these two operations within 50 μs. By considering that both operations run concurrently, separate hardware architectures for calculation and communication are considered.

The calculation architecture is designed based on the multi-compartmental extended HH, where each compartment holds a certain state potential that is updated after every simulation step. The new state of the compartments is dependent on the simulated currents that either enter or leave the compartment, and the coupling effect of the neighbouring IONs. When inspecting the compartments closer, they can be split into two groups. Within the first group are the calculations that are independent of the neuron network topology, which are housed by soma and axon (hillock) compartments. The second group holds calculations that are dependent on the coupling effect of how many neighbours are around the SimC (topology), and consists of the dendrite compartment. By considering that both segments can run concurrently, the following formula gives the total amount of cycles needed to complete the desired calculation:

$$C_{phyc} = max(C_{dend}, C_{a+s}) \tag{3.1}$$

The communication architecture is used to transfer the updated dendrite and axon hillock responses through the network. The neuron network is based on a binary tree network that is connected to several memory controllers. The memory controllers receive and store desired data from the routing network, and through a connectivity matrix, pass neighbouring ION (dendrite) potentials onto the PhyCs, which calculate specific SimCs. The resulting responses from these SimCs are then stored and sent through the tree network to the other memory controllers and the output of the FPGA.

This chapter is organised as follows: Section 3.2 describes the design approach in detail, giving insights into how the design is built up. Sections 3.3 and 3.4 clarify how the proposed design is evaluated and what results are achieved on multiple targets. Section 3.5 concludes the chapter and offers several suggestions for the future work.

3.2 The Calculation Architecture

The calculation architecture carries out the extended HH equations in hardware and is built up in two parts. The first part consists of the grouped topology-independent axon and soma calculations, and the second part is the topology-dependent Dendrite calculation. Both parts of the calculations run concurrently, and next to each other within a PhyC. Furthermore, to reduce the required resources, each or multiple PhyC(s) scheduled around a single-exponent coprocessor.

3.2.1 The Physical Cell Overview

A physical cell is the hardware implementation of the extended Hodgkin–Huxley neuron cell. Each PhyC represents the schedule and floating-point operations that are needed to calculate one ION response excluding the exponent function, which is carried out by the coprocessor. Within each PhyC (Figure 3.1), two concurrently running computing architectures are housed that are activated when a high start signal is given. These represent the combined axon+soma, and dendrite compartment calculations. The axon+soma computational hardware (AS) computes the next axon hillock and soma state and updates a set of cell parameters, based on the current cell compartment states and cell parameters. The dendrite computation hardware (DEND) computes the next dendrite state, first based on the current dendrite state and the coupling effect of the neighbouring dendrite states, and second on a response from the axon+soma calculation. The axon hillock response and dendrite state following these computations are multiplexed by the *OutputMux* operation and given as output of the system or communicated to the other PhyCs.

By iterating multiple computational loops over a single PhyC, several SimC responses can be computed. These SimCs are then said to be time-shared by a certain time-sharing factor (TSF). Many iterations can be computed as long as the total amount of time taken by all iterations is smaller than the time step (T*step*) of 50 μs. When all time-shared SimC responses have

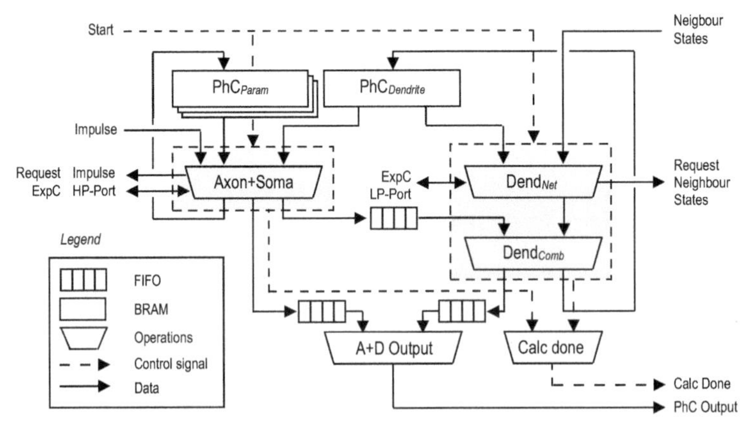

Figure 3.1 Dataflow of a *PhC*. The dashed box on the left is the axon/soma calculation. The one on the right is the dendrite calculation. Cell states are stored in the BRAM (memory). All input and output data signals are connected to FIFO buffers (not shown).

been computed using both the axon+soma and dendrite computational hardware, output "*Calc Done*" signal signifies that all calculations for that part of the PhyC have been carried out. A logical *and* (&) operation is performed on both signals, signifying that the PhyC has finished its computational round, and is waiting for the next T*Step* to begin.

First-in-first-out buffers (FIFO) are used to resolve communication between the concurrently running hardware. As both computations run equally often, and only a single value is sent between the concurrently running parts, each FIFO depth is set to 1, and is seen as a simple buffer. If no computational element stalls during the simulation time, then the system is considered to be matched. In a matched system, the amount of cycles between the axon+soma and dendrite hardware is given by (3.2).

$$|C_{a+s} - C_{dend}| < \frac{min\,(C_{a+s}, C_{dend})}{TSF}. \tag{3.2}$$

3.2.2 Initialising the Physical Cells

Before the PhyC can start generating SimC responses, it needs to be initialised with certain cell parameters. Each PhyC locally stores the parameters and states for the SimCs that it will be generating responses for in-block memories (BRAM). These BRAMs are initialised by the *axon+soma* hardware. After the *axon+soma* has finished initialising the local memory, it sends a signal to the Dendrite hardware signalling that the PhyC has been initiated with valid values. The *axon+soma* hardware requests the initialisation values by sending the PhyC-local SimC address to the cluster controller through an FIFO. The cluster controller then sends the SimC initialisation parameters and states through an FIFO back towards the *axon+soma*.

3.2.3 Axon Hillock + Soma Hardware

The *axon + soma* hardware (AS) generates the axon hillock responses based solely on the cell's current state, and is thus independent of the neural network topology. Scheduling is mostly determined by the Vivado HLS scheduler; however, certain directives are used to inform which hardware components should be chosen, and how to efficiently implement the external exponent coprocessor.

In this section, the *axon+soma* design is explored in three ways. First, the work done to reduce the *axon+soma* complexity and offload some of the *axon+soma* exponent calculations is shown. Second, the HLS scheduler is directed on how to design the computations of the *axon+soma*, and finally, the latency of the resulting *axon+soma* is determined.

3.2.3.1 Exponent operand schedule

When inspecting the 22 exponent operands within the *axon+soma*, it is found that they consist of an addition and multiplication carried out by two constants on a single-cell state (compartmental potential). Furthermore, all equations besides two can be written in the form:

$$e^{\phi},\ \phi = (State + \alpha) \times \beta. \tag{3.3}$$

By rewriting the formula such that ψ_1 and ψ_2 have the same form as (3.3), all exponent operand computations can be carried out by the same piece of hardware using an efficient pipeline. The dataflow diagram for this hardware is shown in Figure 3.2.

The controller for the application-specific coprocessor ensures that the correct state potential and operand constants (α, β) are stored in the registers (State, Op1 and Op2) at the right time. Each result is then sent via an FIFO to the ExpC. To ensure that the Vivado scheduler is informed after what number of cycles from the ExpC are ready, several *wait* states are added. The amount of wait states that are added between every result is equal to the amount of *axon+soma* that are sharing the ExpC. By considering then that multiple PhyCs are scheduled around a single ExpC, and the ExpC can only process a single input every clock cycle, the application-specific coprocessor can be expanded, to be used by all PhyCs reducing the required resources

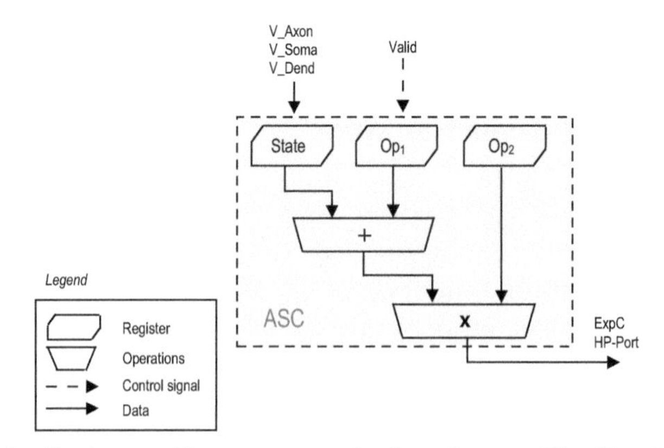

Figure 3.2 Application-specific coprocessor dataflow diagram. The 'State' register holds axon-, soma-, or dendrite-state potential. The 'Op1' and 'Op2' registers hold the constants used for the calculation.

within each *axon+soma*. The application-specific coprocessor is adjusted to receive multiple cell states from multiple sources and to give a source address together with the exponent operand as (addressed) output.

3.2.3.2 Axon hillock and soma compartment controller

The axon hillock and soma compartment controller performs all the operations needed to generate a new axon hillock response and output it. Calculations except those carried out by the application-specific coprocessor are scheduled automatically by the Vivado HLS. In Figure 3.3, the control flow diagram of the application-specific coprocessor is shown. The computation can be split into three segments: Fetch, HLS Scheduler and Output. Before any calculations can be carried out, the hardware needs to *Fetch* the required cell states and parameters from the local memory, while also sending out a request to the memory controller to receive a (non-zero) impulse value. With the fetched values, the *axon+soma* can partly offload its calculations to the application-specific coprocessor, while starting calculations that are independent of the results from the exponent calculations.

The *HLS Scheduler* decides based on when these independent calculations can be computed, which values are retrieved first. Updated values are also stored immediately back in the local memory by the HLS. Lastly, the *axon+soma* outputs an axon hillock in response to the *OutputMux* (Figure 3.1) and a separate in response to the dendrite hardware.

To ensure that the scheduler builds a system that is both fast and resource efficient, certain directives are used. Three types of directives are given (Figure 3.3). First, to reduce the amount of hardware (LUTs and FFs), large arrays are placed in BRAMs. Memory that is needed as soon as possible, such as the state potentials are placed in their own BRAM so they can be accessed

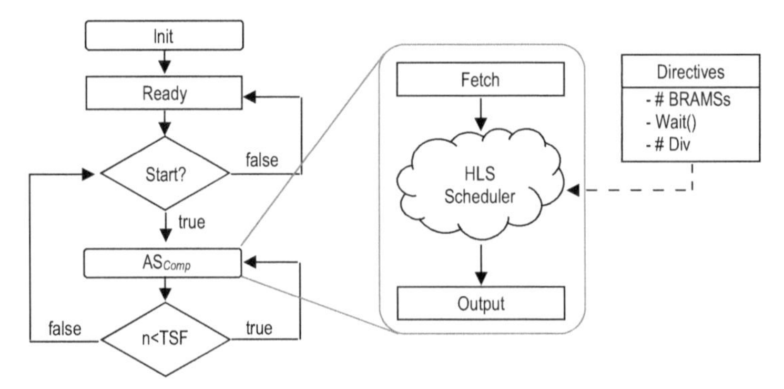

Figure 3.3 Control flow of axon+soma, with the computation flow separately highlighted.

at the same time, while cell parameters are grouped together and can be accessed through a single BRAM port. Second, several wait states are injected into the code to inform the scheduler when a signal outside the scheduler's scope will arrive. By increasing the number of PhyCs that have access to a single ExpC, the amount of cycles between arriving exponent results is increased. Placing wait states equal to the amount of accessing PhyC, between the read actions within the SystemC code, informs the scheduler when to expect a new exponent (resulting) input.

This gives two advantages. First, it gives the scheduler a basic framework, into which it can schedule the C code, resulting in smaller, simpler hardware. Second, due to the given knowledge, no stalls are accrued during the *axon+soma* computation after the first exponent result is read. In addition, it is possible to determine how many floating-point functions are implemented by the HLS of a single function. Besides the exponent calculation, the divider is the largest and fewest used computational function. By instructing the scheduler to design the schedule around a single divider, the amount of required LUT resources is reduced.

For each SimC computed by the PhyC, the *axon+soma* incurs a fixed latency independent of neuron network topology. The latency of the *axon+soma* is determined by the number of PhyCs that share a coprocessor, and how the HLS schedules the axon hillock and soma calculations. It can, however, be affected by two write stalls (FIFO) during the output stage. The two FIFOs in question are the output FIFO to the communication model and the intermediate FIFO to the DEND. These stalls can by mitigated by giving the communication model enough capacity and ensuring that the PhyC is matched.

By building a minimal model with the HLS, the *axon+soma* base latency (C_{AS_base}) is generated. The base latency is independent of the calculation model configuration. A lower bound cycle count, independent of neuron network topology, can be given to the PhyC latency based on these statements:

$$\lfloor C_{AS} \rfloor = C_{AS\_base} + \eta \times (\delta - 1) - \kappa(\delta) \tag{3.4}$$

where η is the number of exponent calculations required by the axon+soma, δ is the maximum number of PhyCs sharing an ExpC and κ is the overlapping factor. By increasing the number of 'wait' statements between incoming exponent, resulting in an ever increasing latency, the overlapping factor reduces the latency by taking into account that the HLS will schedule independent calculations to overlap with the delayed exponent results.

3.2.4 Dendrite Hardware

The *dendrite* hardware architecture (DEND) computes the new dendrite compartmental state based on the current SimC dendrite state, the neighbouring dendrite states and finally an intermittent response generated by the *axon+soma*. In Figure 3.4, an overview is given on the dendrite control flow. After the PhyC has been initialised, the DEND is *Ready* to start a computational round and outputs a *Calc Done* signal. When a start signal is received, the dendrite computation (Dend_{Comp}) can begin and the *calc_done* signal is negated. After the DEND has updated the dendrite state for a certain SimC, the computation repeats for the next SimC, until the DEND has finished computing a number of SimC dendrite states equal to the time-sharing factor (TSF).

Within the Dend_{Comp}, two main operations are carried out. Dend_{Net} computes the coupling effect of neighbouring cells in the network, its input is determined by the current dendrite state that is *Fetched* from the BRAM memory and the neighbouring dendrite states requested from the memory controller. Dend_{Comb} combines the intermittent response received from the *axon+soma* and results from the Dend_{Net}. The resulting dendrite state from this operation is then locally updated and communicated to the memory controller.

The following section will further give an in-depth description on how the two main operations within the DEND are carried out. For the resulting design, the minimum latency is then formulated.

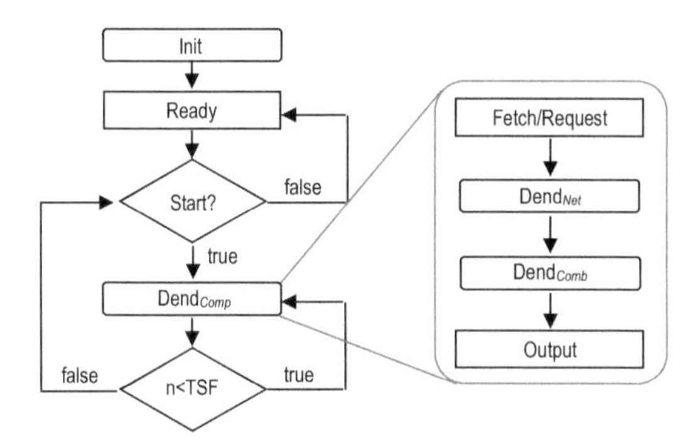

Figure 3.4 Dendrite data flow overview.

3.2.4.1 Dendrite network operation

The dendrite network operation (Dend_{Net}) computes the coupling effect of neighbouring cells in the network. It is scheduled around the exponent operation and is split into two parts: computations that are independent of the exponent result and those that are dependent.

Figure 3.5 shows an overview of the dendrite network operation. After the current dendrite state is *Fetched* and a *Request* is sent to the memory controller, the requested neighbouring state is *Read* as a new token. This token is then passed through the α computation (Figure 3.6(a)), which updates the addition of dendrite voltage differences (V_{Total}) between the SimC and its neighbours and *Writes* the exponent operand (Exp_{Op}) to the ExpC input FIFO. Together with the exponent operand, the difference between the current dendrite state voltage and its neighbour is also written to the V_{Diff} FIFO. This process continues until the V_{Diff} buffer is full and cannot be written to, or all neighbouring states have been received (Token = NaN). In the case of a full buffer, the *Read_Fifo* operation waits until the ExpC has finished at least one exponent operations and feeds this result together with the corresponding voltage difference into the ω computation (Figure 3.6(b)). Afterwards, the buffer can be written again and the control is given back to the previous process. In the case of an NaN token, the Dend_{Net} will carry out the ω computation until the V_{Diff} buffer is empty, signifying all neighbouring states have been computed once over each computations. Afterwards, the results of Dend_{Net} V_{Total} and F_{Total} are combined with the intermittent response from the *axon+soma* by the dendrite combine operation.

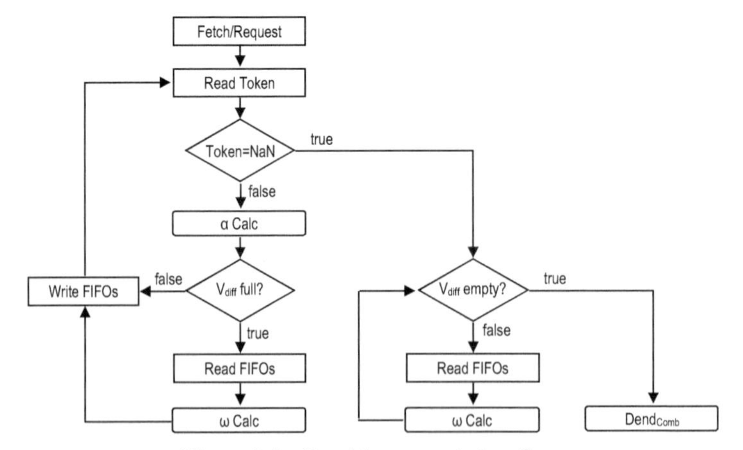

Figure 3.5 Dendrite network data flow.

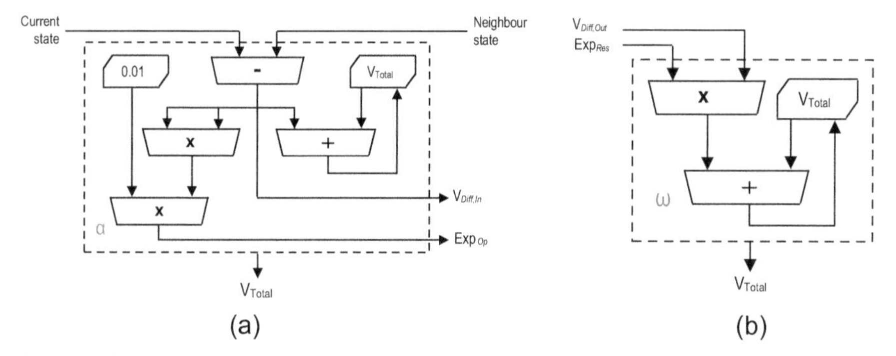

Figure 3.6 The two dataflow diagrams of the dendrite network computations. (a) α computation before the exponent calculation. (b) ω computation after the exponent calculation. $V_{Diff;In}$ and $V_{Diff;Out}$ are the input and output ports to the same FIFO. Exp_{Op} and Exp_{Res} are connected to FIFOs on the ExpC.

3.2.4.2 Dendrite combine operation

The dendrite combine operation ($Dend_{comb}$) combines the results from the $Dend_{Net}$ operation and the intermittent response from the *axon+soma* to generate a new dendrite state potential. Three steps are taken to optimise the original data flow resulting in a reduction of required resources and cycles to complete the computation; a directive is given to limit the amount of implemented multipliers, multiplications with constants are grouped together and, finally, the data path is rearranged.

First, by inserting the directives, the two parallel multiplier operations implemented in the original data path are bound onto a single multiplier. This adjustment reduced the amount of required DSPs by up to 44% per implemented DEND at a cost of one extra clock cycle (pipeline) per SimC computation. Second, the other two multiplier operations can be phased out completely by moving the constant value multiplicands (Δ and (CON)ductance) up the data path and grouping them together with other constant multiplicands. These are then calculated by the pre-processor, resulting in a reduction of the required number of computation cycles. Finally, by rearranging the data flow schedule, the addition of the intermittent signal and the current dendrite state potential that previously took place within the critical path are now moved to take place outside of the critical path (we force the scheduler to retrieve and store the intermittent signal at the start of the data flow). This has again resulted in a reduction of computational cycles.

In Figure 3.7, the timing diagram is shown for the design dendrite combine operation. As all operations are now no longer done in direct succession of each other, two temporary registers are also scheduled.

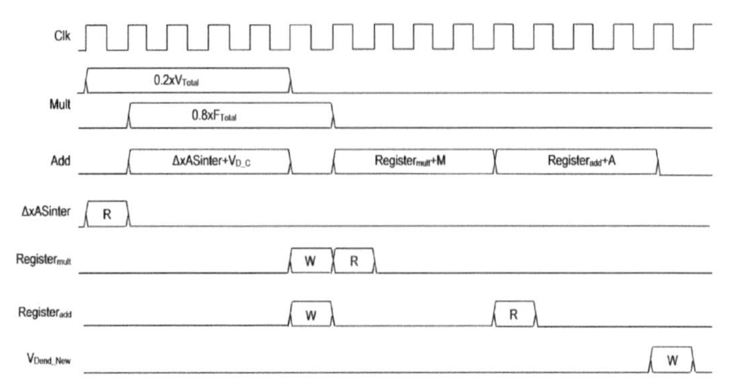

Figure 3.7 Timing diagram of the optimised dendrite combine operation with applied pragma for Virtex 7 double precision implementation. A single multiplier (M) and adder (A) are implemented. The intermittent response, AS_{inter}, is (R)ead from an FIFO, while the output V_{Dend_New} is (W)ritten to the output FIFO and BRAM as the new state of the dendrite.

3.2.4.3 Dendrite compartmental latency

In this sub-section, the lower bound latency to the dendrite compartmental computation (Dend_{comp}) is formulated. The dendrite compartmental latency is split into two parts. The first part consists of the latency attributed to the dendrite network computation, while the second part is the latency caused by the dendrite combine computation.

The dendrite network computation consists of three elements: α computation before the exponent computation, exponent computation and ω computation after the exponent is known. By taking into account that the system is scheduled such that the exponent computation does not get blocked and the V_{Diff} FIFO buffer is large enough, the minimum latency in cycles (C) can be described as:

$$\lfloor C_{net} \rfloor = \begin{cases} C_{net\_base}, & \text{if } N = 0 \\ C_{net\_base} + N \times (C_\alpha + C_\omega), & N \geq M \end{cases}. \tag{3.5}$$

where the base cycles (C_{net_base}) represent the number of cycles needed to request new values from the memory and write the resulting dendrite state back and *N* is the number of neighbours (topology) to the SimC. The C_{net_base} is only counted once because the cluster controller is able to fetch neighbouring dendrite states before α has finished. Note that the latency from the exponent coprocessor does not need to be taken into account when *N* is sufficiently large that

$$C_{ExpC} < (N-1) \times (C_\alpha + C_\omega), \text{ with } N > 1, \tag{3.6}$$

is valid and that the FIFO depth of V_{Diff} is at least the minimal N when 4.5. The latency of the $Dend_{Comb}$ is directly derived from the optimised dataflow. As there are no branches or loops in the $Dend_{Comb}$, the latency (C_{Dend_Comb}) is considered constant. Without taking into account the *axon+soma* latency on when the intermittent value is sent to the DEND, a lower bound of the dendrite compartment latency is formulated:

$$\lfloor C_{DEND} \rfloor = \lfloor C_{net} \rfloor + C_{Dend\_Comb} \tag{3.7}$$

3.2.5 Calculation Architecture Latency

When combining the *axon+soma* and DEND, a minimal latency can be formulated. The *axon+soma* has a higher priority above the DEND when using the ExpC and no direct dependencies, as all input parameters are directly available when a start is given. This makes finding the minimum latency for the *axon+soma* trivial. The DEND, however, is dependent on the *axon+soma* in two cases; First, when an *axon+soma* exponent operand is given to the ExpC, it has priority over a DEND exponent operand and, second, the DEND is dependent on an intermittent response generated by the *axon+soma* (AS_{inter}).

In Figure 3.8, a simple representation is shown on how the PhyC is internally scheduled. Here, the *axon+soma* is split into operational cycles before the intermittent response is known (Pre-AS_{inter}) and after it is given to the DEND (Post-AS_{inter}). The main two components of the $Dend_{net}$ computation are given, followed by the $Dend_{comb}$ operation. Finally, a representation is given for the ExpC; initially, its pipeline is filled with *axon+soma* exponent operands, then *axon+soma* and DEND operand and, finally, only DEND exponent operands.

By increasing the amount of PhyCs that share an ExpC, the ExpC is utilised more effectively, however, at the cost of an extra timing. Also, increasing the amount of neighbours of each SimC will increase the amount of required DEND cycles. Considering that both the final configuration of

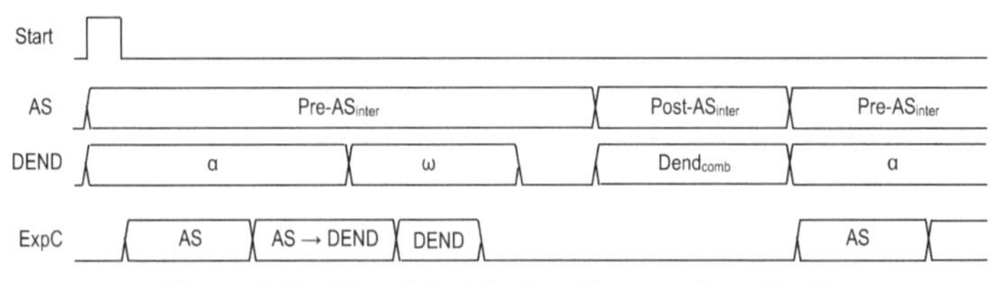

Figure 3.8 Simplified timing diagram for a PhyC.

the calculation architecture and the SimC network topology are yet to be determined, only a minimal latency can be derived:

$$\lfloor C_{Calc\_arch} \rfloor = TSF \times \lfloor C_{AS} \rfloor + max\left(0, C_{DEND_{comb}} - C_{Post-AS_{\text{inter}}}\right) \tag{3.8}$$

3.2.6 Exponent Architecture

In previous implementations [10, 20], the exponent architecture is scheduled and implemented by the Vivado HLS tools. This resulted in several large exponent architectures being implemented within the total hardware topology. When inspecting the amount of times the exponent function is required by the HH algorithm, it is found to be sparsely used; moreover, computing an exponent requires comparatively more resources than other mathematical functions. To solve this problem, a separate controller is developed that forms a bridge between the native SystemC code and an open-source external coprocessor. This controller could then be utilised by multiple PhyCs, resulting in a reduction of required resources at the cost of complexity within the controller.

In Figure 3.9, the dataflow diagram is shown for the exponent controller (ExpC). High-priority inputs originate from the application-specific coprocessor, while the low-priority inputs stem from a dendrite architecture in the PhyC. The '*read scheduler*' acquires new (low priority) input operands based on a round robin schedule with a possible interrupt (high priority). It then feeds each exponent operand and (additional) address into a single channel

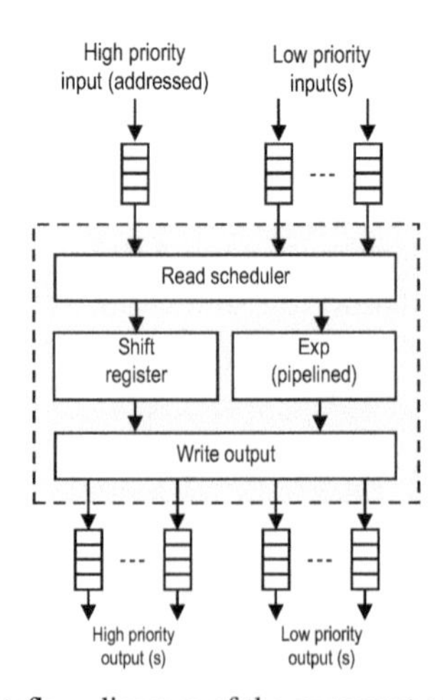

Figure 3.9 Dataflow diagram of the exponent controller (ExpC).

(vector). The vector is then split over pipelined architecture that computes the exponent, and a shift register that keeps track of where the current calculation (address) is in the pipeline. When the exponent is calculated, a valid address is presented at the end of the shift register, signalling a write-back to a specific output FIFO.

To fully integrate the pipelined open-source exponent architecture (FloPoCo) [21, 22], it needs to be configured for the desired implementation target and understand IEEE-754 floating-point types [23]. First, depending on the used FPGA family and chosen clock frequency, configuring the exponent architecture results in a certain pipeline depth (see Table 3.1). The pipeline depth is then also reflected by the length of the shift registers. Second, FloPoCo architectures use a logarithmic number system (LNS) format [24], which is similar to the IEEE standard (used by Vivado HLS SystemC); however, it adds two flag bits to represent the type of number that is given; a natural number, zero, positive or negative infinite and NaN. A simple bridge is designed in hardware to translate these values on the edges of the FloPoCo architecture.

3.3 The Calculation Architecture

The communication architecture is responsible for initialising and delivering requested work to the calculation architecture and communicating to an external interface outside of the proposed design. It is built in three parts on different levels. First, in the lowest level are the cluster controllers that each house local routing tables for a certain set of PhyCs. Second, in the middle level are the routers that pass on values between the cluster controllers and the interface bridge. Finally, in the top level is the interface bridge that connects to the routers and translates the internal signals to an external interface. In this section, these three levels in the communication architecture are discussed.

3.3.1 Communication Architecture Overview

The communication architecture is a reactive element in the design. This means that it is only active when it is given new information to pass on/store

Table 3.1 Exponent latency for different FPGA families with a 100 MHz clock

	Spartan 6	Virtex 6	Virtex 7	Virtex 7 Coregen
32-bit IEEE Float	6	2	2	5
64-bit IEEE Float	14	8	8	8/11[11]

or is requested to retrieve stored memory by the calculation architecture. New data are injected through the top-level interface bridge or on the lower levels by the calculation architecture.

The communication architecture is built to run in two separate modes: initialisation and simulation. Before simulation can start, new parameters are used to initialise the neural network. Dedicated addressable *init* channels are implemented directly from the interface bridge, towards each cluster controller. These channels are used to send the cluster ID, new Hodgkin–Huxley neuron model parameters, cell states or local routing tables. The router logic is however not active while initialising.

After the neural net is initialised, it can be evaluated. While evaluating the network, all calculated dendrite responses are routed through a cluster controller to other cluster controllers (leaves), and the interface bridge (top node). Based on [20], a tree network is a suitable implementation for this routing. Axon hillock responses are only used as an output and thus do not need to be routed to the leaves, only the top node. The routers in the routing network are adjusted accordingly. After each neuron network evaluation step, the system will be idle and wait until it is commanded to evaluate the next step. To communicate that the system is ready to evaluate the next step, a ready signal is generated by a controller that computes the logical conjunction (combine) over the idle signals from the routing network and the cluster controllers.

3.3.2 Cluster Controller

The cluster controller is responsible for relating new values to the calculation architecture when requested, as well as storing and routing their responses while also storing the responses of other PhyCs. Each cluster controller is designed around several parallel running hardware architectures that are synchronised by FIFOs. In Figure 3.10, an example is shown of a cluster controller with two connected PhyCs. The cluster controller consists out of several parallel running hardware controllers.

Init controller: Before the simulation can run, each cluster controller is initialised. Through the initialisation channel, the *Init Controller* receives a coded set of initialisation parameters. To initialise a cluster controller, the following values need to be set: the cluster ID, a local routing table, initiation parameters for the local PhyCs and the dendrite states of all SimCs in the IO.

Write Controller: The write controller communicates with the PhyCs by request. A PhyC only needs to send the job number it is currently working

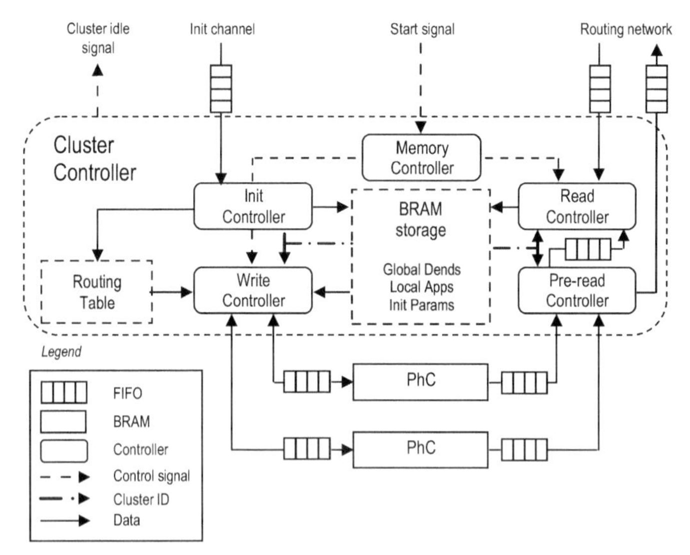

Figure 3.10 Diagram showing how the controllers are housed within the cluster controller; the cluster done logic is excluded from view.

on to inform the write controller what it needs to send. The write controller distinguished two different modes of operation: PhyC initialisation and simulation. When the PhyCs have yet to been initialised and a request arrives, the write controller will transmit the *init* parameters in a certain order starting with the dendrite state. During the simulation, when a dendrite request is received (Figure 3.4), the write controller looks up which neighbouring SimC dendrite states it should send with the help of the local routing table. From then on, the write controller sequential sends each dendrite state addressed on the columns of that row. If an applied current request (Figure 3.3) is received, the local address is used to get the applied current from the BRAM storage.

Routing Table: The neighbouring SimC addresses are placed column-wise in the routing table. Each row represents where a local SimC is located in the IO topology. The row number is determined by comparing by which FIFO channel the request arrives and what job number is sent:

$$\mathrm{Nr}_{Row} = Adr_{Local} = \mathrm{Nr}_{FIFO\_Channel} \times TFS + \mathrm{Nr}_{Job}. \qquad (3.9)$$

Pre-read Controller: After a cell response is generated, the pre-read controller determines the global address of the cell within the neuron network, with the help of the Cluster ID and local SimC response address (3.9). If the cell response is an axon value, then it is sent to the routing network; however, if it is a dendrite response, then the new value is duplicated and sent to both the routing network and the read controller through an internal FIFO for storage.

Read Controller: The read controller is responsible for storing applied currents and dendrite responses in the BRAM storage. New applied currents (i_{Apps}) arrive through the routing network, while dendrite responses can arrive either through the routing network or through the internal FIFO (pre-read controller). To save memory, only applied currents that target the local SimCs are stored. The read controller determines which i_{App}s are admissible by calculating the relevant global address range based on the cluster ID and the amount of SimCs that are connected to the cluster controller.

Memory Controller: The dendrite states and applied currents are stored in the BRAM in two parts: current memory and next-state memory. To inform the write controller that is to read from the current-state memory-block and the read controller to overwrite only the next state memory-block, the memory controller sets a certain signal bit. At the start of each simulation-round, a start signal is received by this controller and the bit is flipped, indicating that the next-state memory block is now current state, and vice versa. This prevents memory being overwritten by the read controller before it can be sent to the PhyC.

Cluster Done Logic: The cluster controller falls into an idle state when the connected PhyCs have finished their calculations, and all newly generated results have been stored and/or sent to the routing network. While in this idle state, the cluster done logic presents a valid signal. It accomplishes this by performing a logic AND operation over all incoming and internal FIFO empty signal pins [19], together with the connected (local) PhyC *Calc_Done* signals.

3.3.3 Routing Network

Information between clusters and the interface bridge is passed through a balanced tree network. The network is based on work by [20], where a certain set of routers transmit information based on pre-configured routing tables. In this section, the routing method and implementation for the proposed design are discussed.

3.3.3.1 Routing method

In a basic IO network, only two types of data are transmitted by the clusters. These are *Dendritic* and *Axon hillock* potentials. While cell dendrite potentials are shared among all IONs, axon hillock potentials are currently only given as an output. On the basis of this rule set, a simple router is designed with the following rules: *i)* in a balanced tree network, each router is connected to one bidirectional upstream and two bidirectional downstream channels; *ii)* new dendrite potential values can arrive through any channel and are passed along the other two channels; *iii)* new axon hillock potential values only arrive through one of the two downstream channels and are then transmitted to the upstream channel and *iv)* the top router is connected to the interface bridge.

3.3.3.2 Design specification

Within the proposed design, each router is connected to three channels and is implemented around a single core (router logic) together with a (FIFO) buffer. The channels consist of an input and output FIFO, forming a bidirectional channel. The router logic reads every channel in a round robin-type fashion. If a new token is present in one of the channels it is read, then based on the type (rule set), the token is transmitted to one or two of the other channels. However, due to hardware limitations, a channel might fill up before it is emptied (read). To prevent the router from possibly stalling, it can temporarily store multiple tokens in the buffer for later transmission. By designing the router around a small FSM, each symbol can be passed to 1 or 2 channels every two clock cycles. Even if both channels are full, the buffer is written twice to, resulting in a small increase in latency of two cycles. To ensure that the network does not stall in certain network areas, and above all that new responses are extracted from the clusters, sufficient memory resources should be allocated to the channels and buffers. Together with an algorithm to automatically generate and connect the network, two algorithms were developed to create enough buffer/channel space.

As all clusters are synchronised by the same start signal, a predictable upper bound of Baud rate can be considered. The Baud rate is based on the time it takes to compute a single response ($\Delta PhyC$) and the total amount of PhyCs that are present in the design:

$$Baud_{Rate} = \frac{2 \times PhyC_{Total}}{\Delta PhyC}. \tag{3.10}$$

Priority is given to the clusters by doubling the size of each channel higher up in the tree. This ensures that there is enough space to store all values from the downstream channels into their upstream counterpart. A recursive algorithm is used to generate and connect the balanced tree network to the clusters and interface bridge.

3.3.4 Interface Bridge

The interface bridge is capable of communicating to and from the (virtual) boundary pins of the design and the routing network. The interface protocol used to communicate to the tree network is implemented as a simple hand-shaking protocol based on a Whishbone [14] equivalent without interrupt. For testing purposes, however, the PhyC responses are streamed to prevent the tree network from stalling. By streaming the generated data from the tree network, a best-case situation is considered where no delay needs to be taken into account for the receiving outside logic.

3.4 Experimental Results

3.4.1 Evaluation Method

Within this section, the evaluation method used to generate and test the ION logic is discussed. The ION logic is configurable for multiple set-ups, each having an effect on logic size (and timing) and calculation time. To evaluate different logic set-ups, multiple test sets are generated. Each test set is used to verify the correct behaviour of the logic compared to the simulated reference model and to assess the required logic resources. By automating this process, large amounts of data can be quickly analysed. Automation can done in batch mode or in single run.

3.4.1.1 Building a test set

A single ION set-up is generated with the help of two different files. The *config file* is used by the C, SystemC and Vivado HLS compilers to determine the ION configuration. The *makefile* configures the config file, runs the code generators, checks to see if the generated code is compilable, runs the reference model, creates a set of Vivado (HLS) scripts and finally moves the generated code into a working directory ready to be tested.

The key configuration options are split into four segments: the ION design configuration, which steps are taken for this test set, how the test set is implemented and finally how the test set is simulated and evaluated.

The ION configuration is configured by the dimension parameters. Dimension three determines how many cluster controllers are implemented and how large the routing network shall be. Dimension one is equal to the amount of PhyCs that are connected to each cluster controller. Dimension two gives the time-sharing factor, and finally, dimension four is used to schedule one or several exponent coprocessors over a specified amount of PhyCs. The total amount of ION neurons is equal to:

$$\#ION_{\ neurons} = DIM3 \times DIM1 \times DIM2. \tag{3.11}$$

The total amount of implemented exponent coprocessor is equal to:

$$\#Exp_{\ coprocessors} = DIM3 \times \lceil DIM1/DIM4 \rceil. \tag{3.12}$$

The second segment tells the toolbox through which steps the test set is run. The third segment informs the toolbox on which FPGA the design will be implemented and which implementation possibilities are available. First, the precision parameter is used to set up the test set with single- or double-precision floating based on [23]. This parameter is not only important to set up, by which precision all calculations are computed, but also on the amount of physical wires within, and on the boundaries of the design. Second, the solution parameter informs the toolbox which IP7 is available to be implemented. Finally, the target part parameter gives limitations to the applicable hardware resources and attainable speed. The last key segment is used to set up how the simulation and evaluation are carried out by the respected tools. The simulation time limits the time that a simulation or evaluation run takes, thereby ensuring that the simulation does not endlessly carries on, preventing log files from overflowing. The number of simulation steps determines how long (time) brain wave patterns are generated. To truly reproduce a stable neuron brain wave, each output value should be communicated within every 50 μs [25]. However, for simulation purposes, the periodic deadline can be scaled up to examine the effects of larger neuron networks on the latency. After the *makefile* has completed the initial set-up, a document structure is created with all the required SystemC and FloPoCo source files together with additional scripts and a testbench that can then be used to simulate and evaluate a fixed design. All parameters that are used within the helper script are copied over by the *makefile* to the config file. This is required because Vivado will read from this specific source file to determine how the fixed ION design is built up.

3.4.1.2 Design simulation

A test set is simulated with the help of Xilinx Vivado HLS. Simulation does not accurately represent the timing behaviour of the final design, but is very useful to 'quickly' test the logic behaviour of the code and find possible syntax errors. As the simulation only takes place in software, the FloPoCo VHDL sources cannot be used. To alleviate this problem, code is explicitly injected at the point where the exponent computation is carried out. This code tells the VivadoHLS compiler to calculate the natural exponent and output its result after a certain number of cycles has passed to match the latency of the coprocessor. When the simulation has finished, a log of all axon hillock responses is compared to the golden reference for that ION configuration.

3.4.1.3 SystemC synthesis

After the behaviour of the source files has been logically compared to the golden reference and no faults are found, the SystemC code is 'synthesised'. With the help of the pragmas, wait statements and the following schedule placed in the source code, the VivadoHLS scheduler creates state and dataflow machines in VHDL. However, to prevent it from trying to implement its own closed-source exponent IP, dummy code is automatically injected back at the spot where the calculation takes place. This dummy code solely consists of a simple shift register that matches the latency of the exponent architecture, resulting in a correct behavioural and latency response when the hardware is evaluated.

3.4.1.4 Post-synthesis simulation

The post-synthesis simulation of the VHDL is carried out in two steps. First, VivadoHLS tries to create a logic evaluation based on the SystemC source. This will fail due to the exchange of dummy code with the an open-source coprocessor, which is currently outside the scope of Xilinx's software tool. Second, the design's logic behaviour (including the coprocessor) is simulated with Modelsim. Here, Modelsim takes the initial input test-bench from the VivadoHLS evaluation, but computes new outputs based on the full design and stores them in a signal wave log. By extracting the axon hillock responses presented as output from this signal wave log, the design logic is validated and the cycle latency is found.

3.4.1.5 VHDL implementation

Finally, to find if the design fits on the desired FPGA, an implementation is created with either ISE or Vivado. By inspecting the size of designs, the effect of the ION configurations on required resources can be extrapolated.

3.4.2 Evaluation Results

3.4.2.1 Accuracy results

The most important facet of the design is that it computes correct responses. Based on an accurate reference model, a 'golden reference' is generated for a particular neuron network size with neurons that have been initialised with all but one set of fixed parameters. The responses of this model are given in double floating-point values. Observed error is very low, less than $0.2 \times 10^{-5}\%$, for cell resting state (when most internal cell variables change rapidly), and at cell firing state, for both 32- and 64-bit configurations.

3.4.2.2 Latency results

The latency of the design depends on the size of the neuron network and how the design is configured. Introducing more neurons to the network would logically require more computational time; however, by also increasing the amount of parallel running hardware, computations can be performed in parallel with only a latency penalty due to communication delay. Below, each ION configuration parameter is explored for the Virtex to find and map the effect towards the overall design latency.

First, the latency differential is explored for increasing the TSF. Increasing the TSF has a nearly linear effect on the design latency, and increasing the amount of computation hardware can strongly reduce the latency up to a certain point where the communication latency between the cluster controller and the PhyCs takes president. Next, the effect of varying the number of cluster controllers is evaluated. We found that scaling up the number of cluster controllers has only a small effect on latency. Finally, the effect of reducing the number of ExpCs is explored. By reducing the number of available ExpCs, while keeping the total number of neurons, TSF and computation hardware fixed, multiple PhyCs will share an ExpC between them and latency is increased. Based on these results, it is shown that by sharing an ExpC over a small amount of PhyCs only has a small effect on latency; however, it quickly scales up. This means that the coprocessor is well utilised.

By extracting the time it takes for one computation, to complete a rough estimate is made on the utilisation factor of the ExpC. Given χ, equal to the total number exponent computations in the network, average utilisation factor per coprocessor is given as:

$$Util_{avr} = \frac{C_{Work}}{C_{Design\ Latency}} \times 100\%, \tag{3.13}$$

$$C_{Work} = \frac{\chi}{DIM3 \times \lceil DIM1/DIM4 \rceil}. \tag{3.14}$$

For the case with eight PhyCs and a single cluster controller design, the utilisation factors are given in Table 3.2. What can be concluded from this table is that the design latency grows faster than the ExpC can keep and that while the utilisation is fairly low, which is a huge improvement over previous that requires five coprocessors per PhyC. This is caused by high-priority operations completely blocking lower-priority operations.

3.4.2.3 Resource usage

Resource usage of an implementation is given as a combination of the four different FPGA resource types: look-up tables (LUTS), ip-ops (FF), digital signal processors (DSP) and block memories (BRAM). By changing the ION configuration parameters, the effect on each of these resource types is explored. Both 32- and 64-bit implementations show that increasing the TSF has little effect on the required resources. Only the number of required BRAMs increases due the extra memory needed to store results of more neurons. Second, the effect of adding more PhyC to the design is inspected. Here, the increase in resources has a direct linear correlation to the amount of implemented PhyC. By introducing more cluster controllers to the design, a tree routing network is created to communicate neuron responses between the clusters. Again adding more hardware to the design increases the amount of required resources; however, unlike with PhyCs, placing more clusters also causes a non-linear amount of routers to be added (e.g. doubling a two-cluster design, triples the number of required routers). When inspecting the required resources, the routers show only to use a minimal amount of additional resources; however, due to the routing complexity on the FPGA, increasing the number of routers causes possible timing violations and eventually becomes un-routable. This effect is especially noticeable with the older six-series FPGAs. Finally, by decreasing the number of exponent coprocessors, less resources should be used. By increasing the ExpC factor, more PhyC are scheduled over a single ExpC. When inspecting the required resources,

Table 3.2 Rough utilisation estimates of an ExpC that is shared among multiple PhyC, for a 64-bit design implemented on a Virtex 7

ExpC Factor	1	2	3	4	5	6	7	8
ExpC Average Utilisation	8,88%	17,77%	23,69%	35,42%	36,68%	32,66%	25,78%	44,94%

there seems to be a decreasing trend when less coprocessors are implemented. Nevertheless, there exist local minima.

3.4.3 Model Configuration

To find an optimal design, first a limit is given to the total amount of implementable physical cells ($\#Tot_{PhyC}$) in the FPGA, based on the critical resources. By dividing this value by the *grouping factor* (φ), the amount of physical cell clusters (*#PCC*) is determined. Results from the previous section have shown that there is an optimum grouping factor dependent on the accuracy (3.17).

$$\#Tot_{PhyC} < \frac{\#Crit\ Resources}{\#\ Resources,\ PhyC} \tag{3.15}$$

$$\#PCC = \left\lfloor \#Tot_{PhyC} \times \frac{1}{\varphi} \right\rfloor \tag{3.16}$$

$$max(\varphi)_{optimum} \leq \begin{cases} 4, \text{when accuracy} = 32\text{-bit} \\ 5, \text{when accuracy} = 64\text{-bit} \end{cases} \tag{3.17}$$

To simulate the biological behaviour of a cell, each physical cell requires a certain number of cycles (C_{phyc}), which is determined by the topology-dependent (C_{dend}) and -independent (C_{a+s}) computation time (in cycles), respectively,

$$C_{phyc} = max\left(C_{dend}, C_{a+s}\right). \tag{3.18}$$

Each neuron response is governed by a combination of the results of the topology-dependent and -independent calculations. By describing the latency of the topology-dependent calculation, C_{phyc} can be written as a function of C_{dend}

$$C_{dend} = max\left(C_{din}, C_{a+s}\right) + \tau. \tag{3.19}$$

The topology-dependent calculation, as a function of the latency of the dendrite calculation to process network inputs (C_{din}), can be split into three sections.

First, we find the start-up delay δ, which is partly dependent on the amount of dendrite calculations that share a memory core. Next, we find the

$$C_{din} = \delta\left(\varphi\right) + max\left(C_{block}, \alpha \times N_D\right) + \omega \times N_D. \tag{3.20}$$

$$N_D = \left\lceil \frac{N}{\#Dend} \right\rceil, \quad \omega \geq \left\lceil \frac{\#PhyC \times \#Dend}{\#ExpC} \right\rceil. \tag{3.21}$$

$$C_{block} \leq \left\lceil 22 \times \frac{\#PhyC}{\#ExpC} \right\rceil + \left\lceil \frac{\#Dend_{Clus} - 1)}{\#ExpC} \right\rceil + \rho, \tag{3.22}$$

$$\#Dend_{Clus} = \#PhyC \ \times \ \#Dend. \tag{3.23}$$

amount of cycles α that take place before each (low-priority) exponent calculation and, finally we calculate the number of cycles ω after the result of the exponent calculation is known.

The calculation of the exponent itself is being carried out by the exponent co-processor with pipeline depth ρ. If the exponent calculation is being blocked by another task after all α calculations are performed, an extra blocking time has to be taken into account (3.22). *#Dend* are the amount of Dend$_{Net}$ operations housed in each physical cell. Increasing the number of exponent coprocessors per physical cell cluster (*#ExpC*) (together with the *#Dend*) with a fixed amount of neighbouring cells (*N*) in (3.20) and (3.22), linearly decreases the C_{din}. However, it also increases the number of required resources; the inverse is also true. By optimising the *#ExpC*, the computation of the model is lower bounded by the topology-independent C_{a+s}. Finally, we have to determine the time sharing factor, i.e. how many times a physical cell will be reused within a given time step (T_{step}). An upper bound for the time sharing factor can be calculated by taking into consideration the communication cycles (C_{com}) needed to send all dendrite and axon potential values through the tree network.

$$TSF < \begin{cases} \frac{C_{step} - C_{com}}{C_{phyc}} & C_{com} < C_{phyc} \\ \frac{C_{step} - C_{phyc}}{C_{com}} & \text{otherwise} \end{cases}, \tag{3.24}$$

$$C_{step} = \frac{T_{step}}{Clk_{period}}. \tag{3.25}$$

The total neuron network size is then given by:

$$NN_{size} = TSF \times \#PCC \times \varphi. \tag{3.26}$$

To find an optimum on all targets Figures 3.11 and 3.12 show how the proposed design scales based on latency and a 100 MHz clock speed.

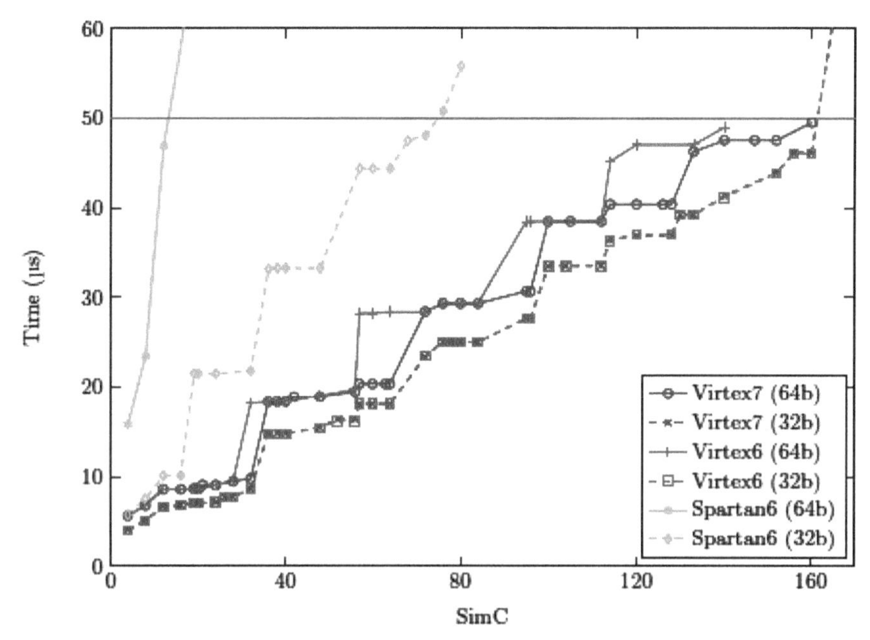

Figure 3.11 Minimum calculation latency for several single-cluster implementable designs. The time step is represented by the straight line at 50 μs.

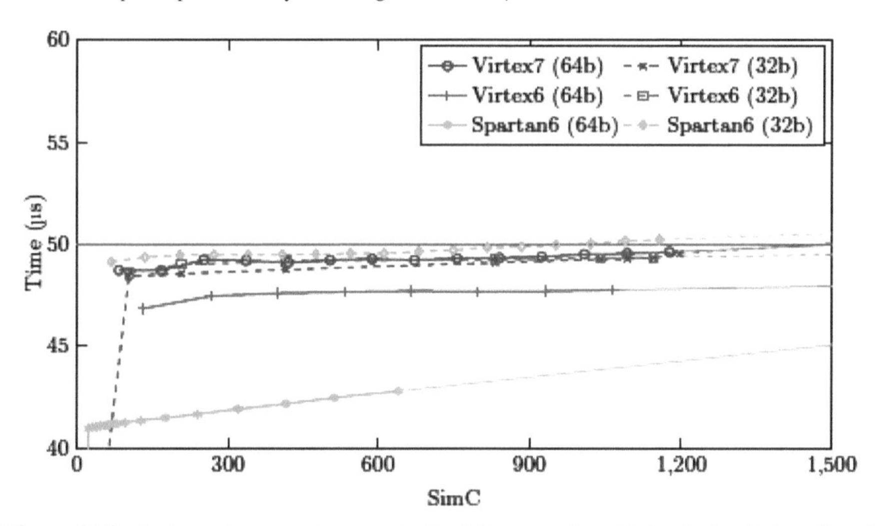

Figure 3.12 Latency increase (communication) for several multiple-cluster designs based on an optimal single-cluster design with simple routing through the tree network. The time step is represented by the straight line at 50 μs.

Table 3.3 Implementable cells on an FPGA with critical resources underlined

Implementation						Resources (Absolute)				Resources (Total Utilisation)				Results	
FPGA	Accuracy	Clusters	PhC	TSF	ExpC	LUT	FF	DSP	BRAM	LUT	FF	DSP	BRAM	SimC[b]	Cost/SimC[c]
[10][a]	64b	NA	8	6	NA	240k	209k	1384	42	69.4%	30.2%	48.1%	1.8%	48	$ 144
Virtex 7	64b	9	2	23	1	324k	202k	1215	233	93.7%	29.2%	42.2%	19.7%	414	$ 15.5
Virtex 6	64b	1	7	20	2	124k	77k	634	122	82.9%	25.8%	82.6%	14.7%	140	$ 20.1
Spartan 6	64b	1	1	8	1	23k	21k	113	27	25.5%	11.6%	62.8%	10.1%	8	$ 29.6
[8][d]	32b	NA	8	12	NA	251k	162k	1600	804	83%	27%	57%	78%	96	$ 51.2
Virtex 7	32b	18	2	33	1	311k	190k	1008	557	90%	27.5%	35%	23.6%	1188	$ 5.4
Virtex 6	32b	4	4	29	2	128k	85k	480	192	85.4%	28.5%	62.5%	23.1%	464	$ 6.1
Spartan 6	32b	1	4	18	2	36k	23k	152	33	39.9%	12.8%	84.4%	12.3%	72	$ 3.3

[a]Only estimates are given in the previous design, built on the same Xilinx Virtex 7 XC7VX550 FPGA board as the current design.

[b]We refer to each (neuron) node in the neuron network as a Simulated Cell (*SimC*), whereas we refer to the hardware used to simulate the cells as a *PhC*.

[c]Reference costs were taken from [27].

[d]Extended HH model, 22.2k operations per neuron in 1 ms, 100% interconnectivity density, 2131.2 MFLOPS.

In Table 3.3, the hardware utilisation for double/single floating-point precision for the main components of the system are shown in terms of flip-flops (FF), time sharing factor (TSF), block RAM (BRAM) and look-up tables (LUT); smaller components like synchronisation circuits are omitted for clarity. All results noted are for the eight-way connectivity inferior-olivary network model.

3.5 Conclusions

ION configuration parameters have an effect on both the latency of the design and the amount of required resources, however each with a different degree. First, the effect of changing the TSF is inspected, then the number of PhyCs, the number of cluster controllers and finally the ExpC factor. By changing the TSF, multiple neuron responses are computed over the same hardware; this has resulted in a large linear effect on the overall latency of the design; however, critical resources are not affected. By changing the number of PhyCs connected to a cluster controller, the size of the routing network does not increase and the cluster controller is better utilised; however, there is only one central memory over multiple PhyCs and they have the possibility of blocking each other. Changing the number of cluster controllers in the design has a slight effect on the latency due to a communication delay between controllers, although it has a larger effect on the usage of critical resources than increasing the number of PhyCs. Finally, by changing the ExpC factor, the design is tweaked to find a satisfying balance between critical resources and speed.

References

[1] H. Gray, *Anatomy of the human body*. Lea & Febiger, 1918.

[2] O. College. (2013, April) Anatomy & Physiology. http://cnx.org/content/col11496/latest/.

[3] E. Izhikevich, "Which model to use for cortical spiking neurons?" *Neural Networks, IEEE Transactions on*, vol. 15, no. 5, pp. 1063–1070, Sept 2004.

[4] J. R. De Gruijl, P. Bazzigaluppi, M. T. G. de Jeu, and C. I. De Zeeuw, "Climbing fiber burst size and olivary sub-threshold oscillations in a network setting," *PLoS Comput Biol*, vol. 8, no. 12, p. e1002814. Available at: http://dx.doi.org/10.1371/journal.pcbi.1002814.

[5] J. P. Welsh and R. Llinas, "Some organizing principles for the control of movement based on olivocerebellar physiology," *Progress in Brain Research*, vol. 114, pp. 449–461, 1997.

[6] W. Gerstner and W. M. Kistler, *Spiking neuron models: Single neurons, populations, plasticity*. Cambridge University press, 2002.

[7] A. L. Hodgkin and A. F. Huxley, "A quantitative description of membrane current and its application to conduction and excitation in nerve," *The Journal of Physiology*, vol. 117, no. 4, pp. 500–544, Aug. 1952.

[8] G. Smaragdos, S. Isaza, M. F. van Eijk, I. Sourdis, and C. Strydis, "Fpga-based biophysically-meaningful modelling of olivocerebellar neurons," in *Proceedings of the 2014 ACM/SIGDA International Symposium on Field-programmable Gate Arrays*, ser. FPGA '14. New York, NY, USA: ACM, pp. 89–98, 2014.

[9] G. Kahn, "The semantics of a simple language for parallel programming," in *Information processing*, J. L. Rosenfeld, Ed. Stockholm, Sweden: North Holland, Amsterdam, pp. 471–475, Aug. 1974.

[10] M. van Eijk, C. Galuzzi, A. Zjajo, G. Smaragdos, C. Strydis, and R. van Leuken, "Esl design of customizable real-time neuron networks," *IEEE, Biomedical Circuits and Systems Conference*, pp. 671–674, 2014.

[11] D. Rodopoulos, G. Chatzikonstantis, A. Pantelopoulos, D. Soudris, C. De Zeeuw, and C. Strydis, "Optimal mapping of inferior olive neuron simulations on the single-chip cloud computer," in *Embedded Computer Systems: Architectures, Modeling, and Simulation (SAMOS XIV), 2014 International Conference on*, pp. 367–374, July 2014.

[12] H. Nguyen, Z. Al-Ars, G. Smaragdos, and C. Strydis, Accelerating complex brain-model simulations on gpu platforms," in *Design, Automation Test in Europe Conference Exhibition*, pp. 974–979, March 2015.

[13] O. S. Initiative *et al.*, "SystemC synthesizable subset draft 1.3," 2009.

[14] OpenCores, Wishbone b3, wishbone system-on-chip (soc)interconnection architecture for portable ip cores."

[15] R. Nane, V. M. Sima, C. Pilato, J. Choi, B. Fort, A. Canis, Y. T. Chen, H. Hsiao, S. Brown, F. Ferrandi, J. Anderson, and K. Bertels, "A survey and evaluation of fpga high-level synthesis tools," *IEEE Transactions on Computer-Aided Design of Integrated Circuits and Systems*, vol. 35, no. 10, pp. 1591–1604, 2016.

[16] *Vivado Design Suite User Guide, High-Level Synthesis*, v2014.1 ed., Xilinx, May 2014.

[17] "IEEE standard for standard systemc language reference manual," *IEEE Std 1666-2011 (Revision of IEEE Std 1666-2005)*, p. 638, Jan. 2012.

[18] F. de Dinechin, M. Joldes, and B. Pasca, "Automatic generation of polynomial- based hardware architectures for function evaluation," in *Application-specific Systems, Architectures and Processors*. IEEE, 2010.
[19] *LogiCORE IP FIFO Generator v11.0*, v11.0 ed., Xilinx, December 2013.
[20] J. Hofmann, C. Galuzzi, A. Zjajo, and R. van Leuken, "Multi-chip dataflow architecture for massive scale biophysically accurate neuron simulation," *Proceedings of International Conference of the IEEE Engineering in Medicine and Biology Society*, pp. 792–795, 2016.
[21] F. de Dinechin and B. Pasca, "Floating-point exponential functions for DSP- enabled FPGAs," in *Field Programmable Technologies*, pp. 110–117, Dec. 2010.
[22] J. Detrey and F. de Dinechin, "Parameterized floating-point logarithm and exponential functions for FPGAs," *Microprocessors and Microsystems, Special Issue on FPGA-based Reconfigurable Computing*, vol. 31, no. 8, pp. 537–545, 2007.
[23] "IEEE standard for floating-point arithmetic," *IEEE Std 754-2008*, p. 70, Aug. 2008.
[24] M. Haselman, M. Beauchamp, A. Wood, S. Hauck, K. Underwood, and K. S. Hemmert, "A comparison of floating point and logarithmic number systems for fpgas," in *13th Annual IEEE Symposium on Field-Programmable Custom Computing Machines (FCCM'05)*, pp. 181–190, April 2005.
[25] P. Bazzigaluppi, J. R. De Gruijl, R. S. Van Der Giessen, S. Khosrovani, C. I. De Zeeuw, and M. T. De Jeu, "Olivary subthreshold oscillations and burst activity revisited," *Frontiers in Neural Circuits*, vol. 6, pp. 1–13, 2012.
[26] N. Schweighofer, K. Doya, and M. Kawato, "Electrophysiological properties of inferior olive neurons: a compartmental model," *Journal of neurophysiology*, vol. 82, no. 2, pp. 804–817, 1999.
[27] AVNET. Avnet express. Accessed on 19 September 2015. [Online]. Available at: http://avnetexpress.avnet.com
[28] B. Torben-Nielsen, I. Segev, and Y. Yarom, "The generation of phase differences and frequency changes in a network model of inferior olive subthreshold oscillations," *PLoS Comput Biol*, vol. 8, no. 7, p. e1002580, 2012.

4

Digital Neuron Cells for Highly Parallel Cognitive Systems

Haipeng Lin

Delft University of Technology, Delft, The Netherlands

The biophysically meaningful neuron models can be used to simulate human brain behaviour. The understanding of neuron behaviours is expected to play a prominent role in fields such as artificial intelligence and treatments of damaged brain. Mostly, the high level of realism of spiking neuron networks and their complexity require sufficient computational resources limiting the size of the realized networks. Consequently, the main challenge in building complex and biologically accurate spiking neuron network is largely set by the high computational and data-transfer demands. In this chapter, several efficient models of the spiking neurons with characteristics such as axon conduction delays and spike timing-dependent plasticity in a real-time data-flow learning network have been described. With the performance analysis, the trade-offs between the biophysical accuracy and computation complexity are defined for the different models. The experimental results indicate that the proposed real-time data-flow learning network architecture allows the capacity of over 1188 (max. 6300, depending on the model complexity) biophysically accurate neurons in a single FPGA device.

4.1 Introduction

Spiking neural networks (SNNs) [1] replicate the dynamic behaviours and information processing mechanisms of a biological neural system [2] and exhibit temporal pattern processing [3] and fault-tolerant capabilities [4]. Subsequently, the main challenge in designing complex and biologically accurate SNNs is primarily set by the high data transfer and computational demands. Nevertheless, it is only through large-scale networks and/or real-time simulation that biological dynamics for specific experiments, e.g. brain–machine interfaces, can suitably be modelled. Execution of such networks on CPUs with generic programming suites or neuromodelling-specific languages, however, requires a prohibitive amount of time to complete. Field-programmable gate arrays (FPGAs), although slower than custom-made ASICs, due to the inherent high parallelism, are capable of providing enough performance for real-time and even hyper-real-time neuron network simulations. Additionally, via (partial) reconfigurations of the hardware, various neuron models, (e.g. Hodgkin–Huxley model [13], integrate-and-fire model [30] and Izhikevich model [28, 29]), as well as different network topologies and cell interconnect schemes can be simulated. This is substantially enhanced by the use of high-level synthesis tools (e.g. Vivado HLS), which speed up the development process.

In this chapter, besides extended Hodgkin–Huxley model described in Chapter 3, two additional models (integrate-and-fire model and Izhikevich model) of the spiking neurons representing different trade-offs between the biophysical accuracy and computation complexity, whose characteristics such as axon conduction delays, spike timing-dependent plasticity (STDP) [21] and electrochemical state descriptions [16], are implemented in a real-time data-flow learning network.

In the system, the input is localized for each neuron cell, i.e. the input signals can be transmitted to the specific neuron cells at the same cycle. Concurrently, the parameters are also localized for each cell, when the neurons are initialized. Additionally, the implemented system offers configurable on- and off-chip communication latencies as well as neuron calculation latencies. With Vivado HLS, (partial) reconfigurations of hardware, as well as different network topologies and cell interconnect schemes are simulated to find the optimal configurations of the models. The experimental results indicate that the proposed network architecture can implement over 1188 (max. 6300, depending on the model complexity) biophysically accurate neurons in a single FPGA device.

The interface is designed to recognize different external inputs offered to each neuron model and to translate these signals from analog to digital. The data in the form of packets are then transmitted over the network. The input is localized for each neuron cell, i.e. multiple packets can be transmitted to the specific neuron cells at the same cycle. Concurrently, the parameters are also localized for each cell. The localization of parameters can better simulate the biological meaningful models in a real circumstance. [15]. Several models of the spiking neurons representing different trade-offs between the biophysical accuracy and computation complexity are implemented in a real-time, data-flow learning network, as well as the characteristics of neuron models such as axon conduction delays, spike timing-dependent plasticity (STDP) and electrochemical state descriptions. Vivado HLS was used to synthesize and validate the new network and implement it on the FPGA. Via (partial) reconfiguration hardware, as well as different network topologies and cell interconnect schemes, the optimal configurations are defined for the neuron models.

This chapter is organized as follows: Section 4.2 introduces related neuron cell models, integrates these models in the network. Section 4.3 presents an overview of the system design and implementation decisions. Section 4.4 presents an evaluation of implementation results. Section 4.5 concludes the chapter based on the experimental results and offers several suggestions for the future work.

4.2 System Design Configuration

This section presents the features of high-level system design. First, requirements of designed system are listed and then corresponding solutions are introduced. Finally, some adjustments are made to implement the designed system on an FPGA board, within time and resource constraints.

4.2.1 Requirements

The designed system should generally simulate multiple neuron models in the same architecture. Each neuron model is designed as an independent module using SystemC, and can be individually selected and plug in the system. Compared with a system with single-neuron models [15, 4], this designed system encounters several important challenges.

- When given input stimuli, the neuron cell starts computing. For each neuron model, the values and frequencies of external stimuli, as well as the types of input data are different. Contrary to the integrate-and-fire model and the Izhikevich model, where the current is given as a stimuli, the Hodgkin–Huxley model receives the data in the form of potential difference.

- Each neuron model consists of a set of parameters. The biological meaning of each set of parameters is quite different. In the Hodgkin–Huxley model, tens of parameters (19 parameters are defined in this system) record various states of neuron, such as membrane potential and activation of Na and K currents. Sets of parameters in the integrate-and-fire model and the Izhikevich model decide the patterns of spike train. Once the value and frequency of external stimuli were defined, a specific set of parameters can result in a spike train according to one of the 20 properties of neuron model [27].
- The characteristics of each neuron model vary significantly; the system should handle differences in the implementation of each model within the same network, while offering the possibilities for the independent modules of computations (e.g. the Izhikevich model should implement axon conduction delay, while zero-conduction delay is set for the extended Hodgkin–Huxley model and the integrate-and-fire model).
- When implementing the system on an FPGA board, resource and time constraints should be taken into consideration. With the tool of Vivado HLS, some directives can be added to optimize the design. Furthermore, localizing the individual design of neuron model as much as possible is a good solution to reduce resource cost, in case redundant parts take up too much hardware resources when switching within three models.

4.2.2 Input and Output

Any potential or current that originates from the outside world and represents the measurements of the physical quantity is an analog signal. Before the stimuli is sent to the neuron cells, it should be converted to digital signal first, which is discrete time signal generated by digital modulation. The interface connecting to external environment in this system can recognize both types of input (potential or current). It only focuses on the value of signal and data format (e.g. fixed point, floating point), recognition of data types is determined manually while simulator switches to one specific neuron model. Once conversion of signal is done, the corresponding digital signal will be assigned with id and transmitted to destination of neuron cell in the network.

In biology, a nerve system receives multiple external stimuli concurrently. In order to be close to the real situation, this interface is capable of localizing input for each neuron cell. In each simulation cycle, the designed interface can deal with multiple external sources and send these packets with id one by one to the destinations. The id in packet is corresponding to a specific neuron cell in system, which is given at the initializing stage. After the packet is sent into the network, routers utilize the look-up routing tables to forward them to the next node. Finally, the cluster containing the destination of neuron cell will accept the packet and store it in the shared memory, so that the corresponding neuron cell can read it before calculation. Additionally, the designed interface is also responsible for changing the results of calculation back into analog signal and writing them to external files at the end of each simulated time cycle. Meanwhile, these results can be locally chosen to store. Once the range of selected neurons is defined in the configuration module, the other packets of result outside the range will be filtrated by the interface.

4.2.3 Parameters

Sets of parameters are locally stored in the memory of independent module for each neuron model. This design has two advantages: i) It is convenient to switch the neuron models in the network; otherwise, the configuration module would be rewritten each time; ii) Since the calculations in the neuron need to read or write parameters at each cycle, localizing storage of parameters can significantly reduce latency, in particular communication cost. For each neuron cell, the parameters are set individually (e.g. a set of parameters represents various states of neuron cell). Before the system starts working, the parameters are initialized by several random functions. Thus, each neuron can have its own states. These functions specify the parameters within some certain ranges for the neurons, whose maximum and minimum values are pre-defined in the configuration file.

4.2.4 Scalability of Network

A popular connection scheme, which promises high performance and efficiency, as well as low resource cost, is network-on-chip (NoC) [35, 17]. A binary tree is selected to implement the network, since the tree structure can be scaled up or down easily. In the designed network, leaf nodes represent clusters, which contain multiple PhCs inside, while the other intermediate nodes are routers. Each router consists of three input and three output ports. One input (output) port is upstream in the tree, while the others are downstream. Routing tables are attached to each port, so that routers can forward the receiving packets and avoid sending back the packets to loop. Additionally, several FIFO buffers are designed in the router to handle congestion. Each input or output port has corresponding write or read buffers. Once write buffers are full, the following packets will be stored in the delayed buffer first. As soon as the write buffer is empty, the packets in the delayed buffer will be picked up and forwarded to destinations via write buffer. In order to simplify the design, all packets are defined with the same size and priority in the network. Hence, there is no need to split packets into several flits.

In this chapter, the amount of simulated neurons is dependent on three dimensions: the number of clusters, the amount of PhyCs in one cluster and the time-shared factor (the number of neurons allowed to work together in one PhyC). These parameters are all pre-defined in configuration file. Once system starts, it first initializes the network based on the parameters of configuration. A recursive function is designed to generate new branch from root node, until the number of leaf nodes is equal to or larger than the pre-defined parameter. Consequently, by changing the parameters in the configuration file, the network can scale up or down conveniently.

4.2.5 Neuron Models Implementations

Implementation of the extended Hodgkin–Huxley model is based on [4, 15]. External inputs and parameters are localized to achieve the overall goals of system design. This means that multiple inputs can be accepted at each cycle and sent to the destinations individually. Additionally, each neuron cell is initialized with a set of parameters, where the values are different within some specific ranges. Calculation of the Hodgkin–Huxley model can be separated into two parts [16]. The first half is computing potential of dendrite and internal current. At each simulated time step, component of dendrite will first access the share memory in a cluster, to read the potentials from its neighbours and from the external source (Vi). Once all these potentials are calculated, the potential difference can be used to generate the current as:

$$I_C = C * \left(0.8 * \sum_{i=o}^{N-1} (V_i * e^{\frac{-1*V_1^2}{100}}) + 0.2 * \sum_{i=0}^{N-1} * V_i \right) \tag{4.1}$$

where C is the conductance and N is the number of connected neighbours. On the other hand, as long as receiving current via internal channel, the component of axon will generate a new potential of axon and update the parameters. At the end of cycle, new potential of dendrite will be stored in the local memory, and the axon potential will be sent upstream into network via the module of cluster.

Compared with the Hodgkin–Huxley model, the computation of the integrate-and-fire model is simpler. The mathematical expression is below:

$$v_{new} = I + a - bv, \text{ if } v \geq v_{thresh}, \text{ then } v \leftarrow c \quad (4.2)$$

where v is the membrane potential, I is the input current and a, b, c and v_{thresh} are parameters. When the membrane potential v reaches the threshold value v_{thresh}, the neuron will fire a spike, and v is reset to c.

A single neuron of the integrate-and-fire model is designed to connect with $(n + m)$ nodes, where n is the amount of neurons sending the responses to the specific cell via the pre-synapse and m are the ones receiving responses from the neuron via post-synapse. Both n and m are random values defined by a connecting matrix. This matrix is automatically generated using Python [37] and loaded by the system as a script (JSON) [20], when the simulator starts to work. Hence, in each cycle, a neuron of the integrate-and-fire model will detect whether there are some spikes coming from its neighbours or external sources first. Then, a new input current can be calculated, depending on the receiving spikes and external signals. Afterwards, this current will be taken into the function (4.2), resulting in a new potential of dendrite. If this new potential exceeded specific threshold, soma will generate a spike and pass it to all relative neighbours via post-synapses. Concurrently, the potential of dendrite is reset to a constant, which is specified by the parameter in the configuration file. On the contrary, if the new potential of dendrite cannot satisfy the requirement of generating a spike, it will be stored locally in the memory, waiting for the next round of calculation.

The Izhikevich model is expressed as

$$v = 0.04v^2 + 5v + 140 - u + I, \text{ if } v \geq 30, \text{ then } v \leftarrow c, \ u \leftarrow u + d \quad (4.3)$$

$$u = a(bv - u) \quad (4.4)$$

where v represents the membrane potential of neuron and u represents a membrane recovery variable, which account for the activation and inactivation of K+ and Na+ ionic currents, respectively. Besides the characteristic of spikes, the neuron of the Izhikevich model should also implement axon conduction delay and STDP [29]. In this work, the delay between two neighbours is randomly assigned from 0 to a pre-defined max value, and is stored in the local memory (local delay information). Because the generation of spikes and transmission to the neighbours are performed within the same cycle, the spikes will be stored in the delay buffers first. Subsequently, they are remarked with actual arriving time, according to the local delay information. In each cycle, the neuron checks whether there are some spikes in delay buffer meeting the real injecting time. Hence, all suitable spikes will be fetched and used in spiking functions above.

Finally, the system adds a global memory in a top module to record and control synaptic weights for all neuron cells. This design is aimed to realize dynamic change of synaptic weights among neurons real-time conveniently. When a neuron generates a spike, it will store its own STDP parameters in the global memory and notify the neighbours. When the neuron is informed that some neighbours generate spikes, it will update the STDP parameters in the memory, in accordance with the STDP functions. In the Izhikevich model, the synaptic weights directly influence the amount of the current flow through the synapses, which is used to calculate the potential of dendrite. Alternatively, if no spike is fired between two cells, the relative synaptic weight will decay proportionally.

4.2.6 Synthesis

To increase the efficiency, several adjustments are made in the design: i) Arrays are placed in BRAMs to reduce the cost of hardware (LTUs and FFs). Arrays which should be accessed in parallel are separately placed in their own memory, while some other data, e.g. parameters, are assigned to the same BRAM sharing a signal memory port; ii) Since several PhyCs share the same memory in a cluster, wait states are needed in the SystemC code so that no stalls occur during the computations of neuron cells; iii) The division calculation is rarely used since the unit of divider takes up quite a lot of hardware resource. By instructing scheduler to limit the amount of dividers, multiple neuron cells can share one single divider to perform the dividing functions, reducing hardware cost; iv) In the general network, the hardware resource consumed by the design of individual neuron model will become redundant, when the system is switched to another model. Consequently, implementing characteristics of each neuron model, as much as possible, in the independent module, can significantly decrease the overall resource utilization in the system; v) Either single or double floating-point precision can be chosen in the system. The extended Hodgkin–Huxley model involves multiple exponential computations. To save the resource costs, a particular sub-module named ExpC is designed to perform the exponential calculations [4]. In this chapter, several PhyCs of the same cluster can share one ExpC. The number of ExpCs in one cluster is determined by the time-share factor, which is pre-defined in the configuration files. In each cycle, the neurons send the data, including exponent operands and their own addresses, to the corresponding ExpC in pipeline. Once the ExpC received the data, the exponent operands and address are stored separately in two FIFO buffers. After an exponent is calculated, the ExpC will read the address at the top of address buffer first. Subsequently, the data with specific address will be written back to the corresponding neuron via output buffer. As there is only one exponential calculation in the integrate-and-fire model and none in the Izhikevich model, the sub-module of ExpC is removed in their independent modules to reduce the hardware costs. Instead, some additional resource utilization is required to perform the characteristics of delays, and STDP in the Izhikevich model, e.g. some more memories are required to store the delayed spikes, delay information and synaptic weights. This optimized system can find a suitable trade-off between the performance and resource cost. Using Vivado HLS, the FPGA net list is finally generated automatically.

4.3 System Design Implementation

4.3.1 Interface

In this system, the interface is designed to read the signals from external sources and write the simulation results back into the specific files. This input/output module consists of three sub-modules: ADC, DAC and Inout.

4.3.1.1 Inputs and outputs

The ADC is responsible for conversion of the arriving analog signals into digital signal. The DAC writes the results back to the specific external sources. When the module Inout receives data from the ADCs, it assigns the data with a unique id starting from $n + 2$ (n is the number of neuron cells in the system, and the id with n or $n + 1$ is set for the control of synchronization). Afterwards, the data are sent into network in the form of packet, with the same size and priority. On the other side, the packet containing result from network is extracted in the Inout. According to packet id, this result will be written back to the specific external address.

4.3.1.2 Locality of data

4.3.1.2.1 *Localization of inputs*

Multiple external signals can be sent to specific neurons in each cycle. A matrix of connectivity is defined before system starts, based on the parameters in configuration file. This matrix is a two-dimensional ($N \times N$) array ($N = n_c + n_{input} + n_{output} + 2$, where 2 is a constant for the control of synchronization, n_c is the number of neuron cells and n_{input} and n_{output} represent the amount of external inputs and outputs, respectively). An example of matrix is given below:

$$a_{ij} = \begin{bmatrix} 0 & 0 & 0 & 1 & 1 \\ 0 & 0 & 1 & 0 & 0 \\ 0 & 1 & 0 & 0 & 0 \\ 0 & 0 & 1 & 0 & 1 \\ 0 & 0 & 1 & 1 & 0 \end{bmatrix} \tag{4.5}$$

The condition $a_{ij} = 1$ means that the neuron i links to the neuron j, and $a_{ij} = 0$ stands for no connection between the corresponding two neuron cells. When initializing the network, the construction of routing tables is based on the matrix of connectivity. The range of id from $n_c + 2$ to $n_c + 2 + n_{inputs} - 1$ is assigned to the external inputs, each one is relative to a specific input address. By setting the value to 1(0) in the matrix, a pair of input and neuron can be connected (disconnected). Hence, the system is capable of specifying each input for the required neurons. The id n_c or $n_c + 1$ is reserved for the control of synchronization. At the beginning of each cycle, the module of controller (control bus) will send out the packets containing the id n_c to all clusters. Once the clusters received these packets, they will issue the new starting signals to their own neuron cells immediately. On the other side, the packets with id $n_c + 1$ are used to notify the controller that all clusters have already finished calculations, which are generated by each cluster. A count register is placed in the controller, in order to account for the number of arriving packets with id $n_c + 1$. As long as this count equals the number of clusters, the controller will give out the new iteration command in the form of packet (id = n_c).

4.3.1.2.2 *Localization of outputs*

The localization of outputs allows selective observation of specific cells. The id starting with $n_c + 2 + n_{inputs}$ belongs to the external outputs. When setting the value to 1(0) for a pair of neuron and output address in the matrix, the simulation result from the neuron can (not) be written back to the corresponding output files via network.

4.3.2 Implementation of the Neuron Models

4.3.2.1 The extended Hodgkin–Huxley model

4.3.2.1.1 *Neuron cell*

The data flow of the extended Hodgkin–Huxley model is illustrated in Figure 4.1. Before the calculations start, the parameters are initialized by some random functions, where the values are specified within defined ranges. Concurrently, the value of initial V_{Dend} (potential of dendrite) is given via control bus, reading from the configuration file. Due to parameters and V_{Dend} accessed frequently during the computations, both of them are stored in the local memory individually.

The inputs for a single neuron involve two parts: the potential from the neighbours and the external stimuli. The cluster receives the inputs via the network and records them in the share memory at each time step. Then, the neuron cells in the cluster can access the share memory to get their own inputs in the Round-robin order. The computation of the Hodgkin–Huxley model can be split into two sub-computations. The first one can calculate the new V_{Axon} (potential of axon) and the states, based on the current V_{Dend} and parameters in the local memory. The other one, generation of new V_{Dend},

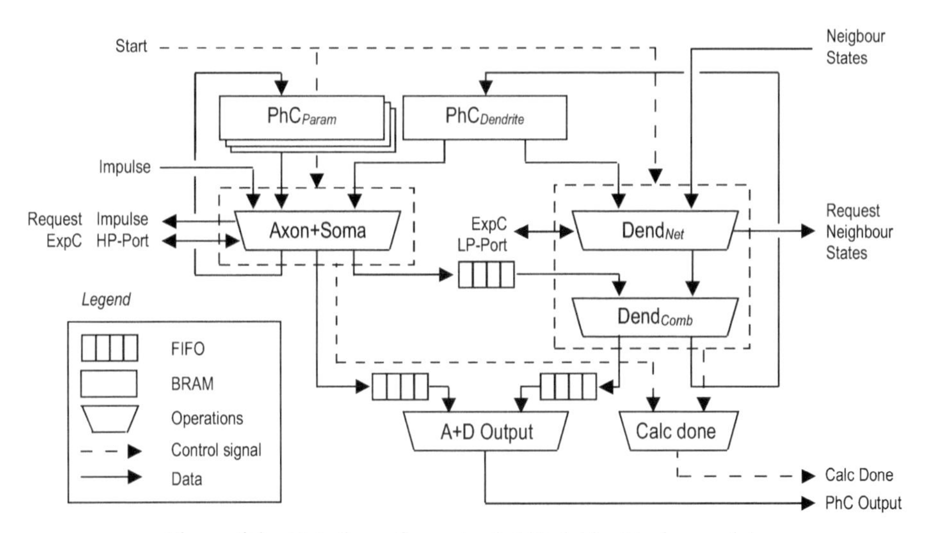

Figure 4.1 Data flow of an extended Hodgkin–Huxley model.

should wait for the new V_{Axon} to continue with the calculation, together with inputs and current V_{Dend}. After the calculation is performed, this new V_{Dend} will be transmitted to the neighbours and V_{Axon} is sent upstream directly into the network, which will be stored in the output files.

4.3.2.1.2 *Physical cell*

Physical cell (PhyC) calculates multiple time-shared neuron cells sequentially [6]. The structure of PhyC has been described in Chapter 3.

4.3.2.1.3 *Cluster*

A single cluster in the system consists of several PhyCs, a share memory and a controller. A simple structure of a cluster has been shown elsewhere. In one cluster, several PhyCs are grouped, which share one memory. At each step, the PhyC accesses this shared memory via a single memory port in the Round-robin order. The shared memory is split into two sub-memories. All writes are done to one memory, while all reads are done from the other memory. After one system step, these two memories exchange their tasks.

The controller in the cluster is designed to implement several functions: i) The controller receives the packets and stores them in the shared memory, according to the address contained in the packets. Meanwhile, the packets from PhyCs are sent upstream to the network by the controller; ii) When receiving a packet with id n_c, the controller will send a start signal to all PhyCs. On the other hand, when the controller received all calculation *done* signals from its PhyCs, it will issue a done packet (id = n_c + 1) to the system controller; iii) The controller can check whether all required packets are received or sent out already at each step.

4.3.2.2 Integrate-and-fire model

A data flow of the integrate–and-fire model is shown in Figure 4.2. The inputs are in the form of current, consisting of spikes from the neighbours, and the external stimuli. All the inputs are added together to derive a total current. This current is used in the potential function to calculate a new V (membrane potential), together with parameters and current membrane potential. Afterwards, the parameters are updated in the memory, and the new V is compared with a defined threshold.

If V exceeds it, a new spike will be generated and transmitted to the neighbours via the network; meanwhile, V and u will be reset and stored in the local memory, respectively. In contrast, if no spike is fired, the new V will be written back to its local memory, waiting for the next step.

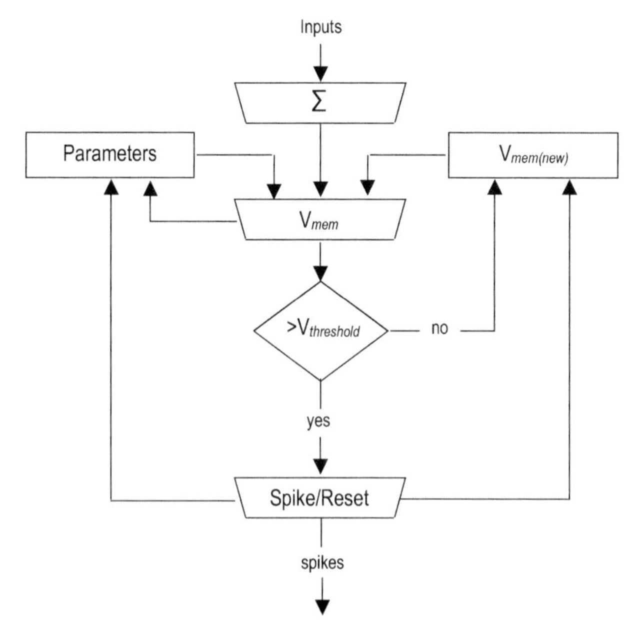

Figure 4.2 Data flow of the integrate-and-fire model.

```
1 V_dend_tmp = V_dend_buf[i]; // read  the current potential of
                             // dendrite  in the memory for  the
                            // neuron i
V_inf_tmp=E_L_tmp + I_tmp*R_m_tmp; //  after the potential of
                                  // dendrite exponentially
                                 // decaying, the temporary
                                // potential V_inf_tmp is
                               // calculated. E_L_tmp, R_m_tmp
                              // are the parameters  read from
                             // the local memory, I_tmp is the
                            // input current of neuron i.
V_dend_tmp = V_inf_tmp +(V_dend_tmp-V_inf_tmp)* exp_fun((0.0-
  dt_tmp)/ tau_tmp); // the new current potential is calculated.

if (V_dend_tmp > -55.0)  // compared the potential with the
                            threshold
{
1 V_dend_tmp=V_reset_tmp;
  Axon_tnp=20.0; // neuron i fire a spike, the dendrite potential
                // is reset.
} else{
```

```
Axon_tmp=V_dend_tmp;      // potential  of  axon
}
// Store the new potential of dendrite to local memory
slice_V_dend[i]=V_dend_tmp;
```

4.3.2.3 Izhikevich model

The Izhikevich model is usually employed to examine the spike trains patterns. In this system, provisions are made to dynamically changes synaptic weights. The design of the Izhikevich model can be sub-divided into three parts: spike generation, delays and STDP.

4.3.2.3.1 *Axonal conduction delay*

The implementation of axonal conduction delay includes three basic functions: f_{delay} (defines the exact delay time for each arriving spike), f_{check} (check whether spikes in delay buffer are compatible with simulation time step) and $f_{current}$ (generates the current). At each step, all arriving inputs from neighbours should be compared first. Only the input, whose potential is larger than the defined threshold, is recognized as a spike from neighbour. The delay information is stored in a local memory, which records the exact delay time for each neighbour. If the spikes are accepted, then f_{delay} accompanies these spikes with delay information and then stores them in a delay buffer. Next, according to the current system simulation time, f_{check} checks whether there are some spikes to be fetched out of the buffer. The total input current can be calculated in f_{check}, together with the external signal. The basic data flow is presented in Figure 4.3.

In the designed network, the generation and transmission of packets are performed in the same cycle. Consequently, to implement the delay between neurons, an additional spike delay buffer is placed in the module, which enables that the spikes can be received at the expected time. It is assumed that the currents of spikes from neighbours depend on the pre-synapse weights; thus, the delayed spike buffer only needs to record the actual arriving time for each pre-synapse neighbour. When the spikes are picked from buffer, the corresponding input currents are calculated based on their synaptic weights. The basic delay implementation in SystemC is shown below:

```
if (v_tmp==35)// check whether the receiving signal is a spike
{
D=(*iter).delay[neighbours. ID]+t-1;   // calculate the actual
time step, according to delay information. t is the current
```

```
          //  system  time step.

(*iter).fire[D].neighbours [neighbours.ID]=1; //Store the
        //spike in delay buffer. At time step D, spike of
       //neighbor.ID  can be fetched.
}

  for(int i=0; i<config.getCellCount(); i++)
    {
 if((*iter).fire[t].neighbours [i]>0) // whether spike from
    //neuron i can  be fetched at time t
 {
 slice [(*iter). Infoli_ID]. I_c +=Sweight[i].s[(*iter).ID];
     // with synaptic weight, the current of spike i is
    // calculated. Then  it is added to the total input
   // current.

 }
```

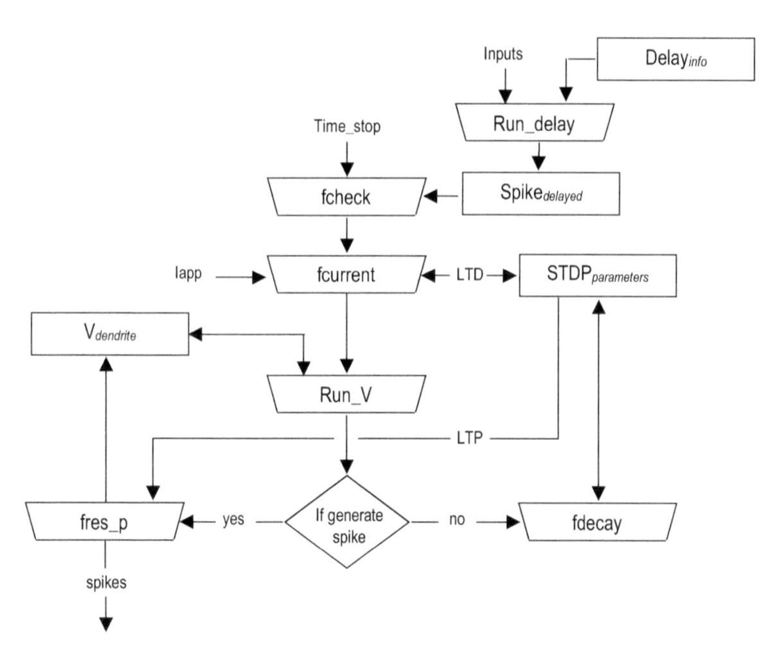

Figure 4.3 Data flow of an Izhikevich model.

4.3.2.3.2 STDP

The implementation of STDP refers to several important variables, e.g. long-term potentiation (LTP), long-term depression (LTD), pre-synapse weight, pre-synapse derivatives, post-synapse weight and post-synapse derivatives. A global memory is employed for the general learning network, to perform the dynamic change of parameters of STDP. When a single neuron of the Izhikevich model receives the spike from a pre-synapse neighbour, the pre-synapse derivative will be depressed by the variable LTD. Next, it updates the new value of STDP parameter in the global memory, so that the neuron firing a spike can know the change of its post-synapse derivative immediately (the pre-synapse weight/derivatives of former equals the post-synapse weight/derivatives of the latter). In the case that a spike is generated by the neuron itself, the variables LTD and LTP will be reset. Furthermore, all the post-synaptic weights of spiking neurons will be increased. These changes are also written into the global memory, so that the post-synapse neighbours can share them. On the other hand, if no spikes generated, all six variables will decay according to STDP rule. The example of STDP implementation in SystemC is shown below:

```
  if(slice[i].V_dend>=30)   // check whether neuron i fire a
                           // spike
{
  Sweight[slice[i].ID].LTD=0.12;
  Sweight[slice[i].ID].LTP[t+config.Max_Delay]=0.1; //reset
                        //LTP and LTD

  for(int k=0;k<config.getCellCount();++k)
  {
  if(neighbour_list[k]. slices[slice[i]. Infoli_ID]) {
     //check  whether the cell k is a neighbor of  neuron  i.
  Sweight[k].sd[slice[i].ID]+=Sweight[k].LTP[t+config.
  Max_Delay-slice[i].delay[k] -1];
    // update  the  synaptic  weights  of  neuron  k
    }
}
  Sweight[slice[i].ID].LTD=0.95*Sweight[slice[i].ID].LTD;
  Sweight[slice[i].ID].LTP[t+config.Max_Delay+1]=
  0.95*Sweight[slice[i].
    ID].LTP[t+config.Max_Delay]; //if neuron i not fire a
                              //spike, the parameters decay,
                             //according to STDP rule.
```

4.3.2.3.3 *Spike generation*

In this designed module, the total current is calculated first, depending on the external inputs, pre-synaptic weight and delays. Then, f_V generates the new membrane potential based on the calculated current, current V_{dend} and the parameters. Both the new potential and parameters are written to the local memory. Afterwards, this new membrane potential is compared with defined threshold to determine whether a new spike can fire. No spike generation means that the parameters of STDP should be decayed proportionally by the f_{decay}. Otherwise, the membrane potential and some specific parameters are reset to the pre-defined values and fed back to the corresponding local memories. Concurrently, all the post-synaptic weights of spiking neurons are increased. The spike generation in SystemC can be described as:

```
slice[i].V_dend=slice[i]. V_dend+slice[i].tau*(0.04*slice[i].
    V_dend*slice[i].V_dend+5*slice[i].V_dend+140-Slice[i].
    u+Slice[i].I_c); //calculating the potential of dendrite
   //for neuran i

slice [i].u +=slice [i]. tau*slice[i]. a*(slice[i].b*slice[i].
   V_dend-slice [i].u);
// update the parameter u in the memory

if(slice[i] .V_dend>=30) // whether neuron i generate a spike
  {
   slice[i].V_dend=-65; // reset potential of dendrite
   slice[i].u +=slice[i].d; // update the parameter u
   slica[i].V_axon=35; // reset potential of axon
  }else
  { slice[i].V_axon=slice[i].V_dend;}  //no spike generated
```

4.3.3 High-level Synthesis

When implementing the designed network on the FPGA, the resource requirements and time constraints must be taken into consideration. To ensure that the synthesis of the system fulfils the requirements, several adjustments are made in the system.

4.3.3.1 Optimization with directives

Using Vivado HLS to synthesize the system, various settings can be conveniently defined by the users. Some main directives used in the system are listed below.

- During the implementation of three neuron models, calculation of divider is the least amount of times used, which consumes considerable hardware resources.
- An appreciate placement of arrays in the memories can effectively reduce the hardware cost. According to the requirements, some arrays (e.g. array storing potential of dendrite) are allocated to the local memories individually, which need to be operated concurrently. Conversely, other arrays (i.e. array on the state parameters) can share the same local memory (neurons in the same PhyC perform their calculations sequentially).
- Several wait() statements are inserted into the code to notify the scheduler of the arriving of signals outside the scope (e.g. the PhyCs access the shared memory in the cluster in the Round-robin order). By instructing wait statements, scheduler can schedule the access to memory for specific PhyC at each cycle, to ensure that no stalls occur and that each one can read or write the shared memory in time.
- The calculations of neuron models support both double and single floating-point precisions.

4.3.3.2 Adjustments of system for HLS

4.3.3.2.1 *Hodgkin–Huxley model*

The implementation of the extended Hodgkin–Huxley model includes multiple computations of exponential functions. If the neuron cells directly make use of the units of exponential function to perform the computation, the resource requirement will increase significantly with the growth of the amount of neurons simulated. Consequently, a special component of exponential calculation is designed [4], which can be shared by the PhyCs in the cluster. The number of these components is not just limited by only one in the cluster. Instead, it is up to the integer value of PhyCs/ExpCs (PhyCs stands for the number of PhyCs in one cluster, and ExpCs means the maximum number of PhyCs supported by one exponential component).

As described in the previous section, the computations of the Hodgkin–Huxley model can be divided into two sub-calculations. The first one is the axon potential and state parameters computations, which includes multiple exponential calculations. The inputs to this part are named low-priority inputs and are sent to multiple ports of exponent core. The second sub-calculation, the generation of dendrite potential only needs one exponential operation. The exponential operand is read by a single port, which is shared with multiple PhyCs. When the exponential operands arrive at the ExpC, the operations are performed in pipeline. A shift counter is an FIFO buffer and placed to record the input address. Once the new input is read, its address will be inserted to the bottom of buffer. While an exponential result is generated at each cycle, the address on the top of the counter will be fetched. Thus, these data can be written back to the specific PhyC via the output buffer. Finally, all the entries in the counter shift up one place and then the bottom of entry is available to the next input.

4.3.3.2.2 *Integrate-and-fire model*

Unlike the Hodgkin–Huxley model, the integrate-and-fire model only needs one exponential operation in the computations. Consequently, the design of this model includes two options: either using the exponential component as described above or directly performing the operation on the unit of exponential function. The latter costs approximately 2.5$\times$ less hardware resource than the former.

4.3.3.2.3 *Izhikevich model*

The computations of the Izhikevich model do not include any exponential operation (consequently, approximately 10% of memory is saved). The STDP parameters are localized in each PhyC. In comparison to the network-saving STDP parameters in the global memory, this adjustment reduces the memory latency. The main problem encountered when implementing this model is how to inform the neighbours of spiking neuron when the spike is fired. This means that the synaptic weight between two neurons should be kept consistent and potentiation or depression of synaptic weights needs to be scheduled to occur at the right time. In comparison to the original matrix recording the connections from the pre-synapse neighbour (e.g. a_{ij} in the matrix means that node a_i connects to the a_j), the new matrix stores the information of post-synapse neuron. In connectivity matrix, each neuron cell is both the pre-synapse and post-synapse neighbour for the connecting node (e.g. if a_i connects to the node a_j, then there must be an inverse connection from the node a_j to a_i). After both matrices are defined, their values are sent to the cluster memory in the form of packets.

At each step, all the dendrite potentials are updated in the local memories for the neurons. When a new cycle of calculation starts, the cluster reads the required data in the shared memory and transmits them to the corresponding neurons via FIFO buffers. The potentials from pre-synapse nodes are sent first. After the cell receives a packet with special value (e.g. less than zero), it is informed that the following packets are from the post-synapse neurons. In the module of Izhikevich neuron, a counter is placed to distinguish potentials. If one input exceeds the pre-defined threshold, the corresponding pre- or post-synapse neighbour fires a spike. Thus, the pre- or post-synapse weight should be increased or decreased (depending on the delay information for this pair of nodes), which are updated in the local memory. In a neuron itself, if the new membrane potential is smaller than the threshold, the synaptic weights will decay according to STDP rule. On the other hand, if a spike is fired, the synaptic weights for all the neighbours are enhanced or reduced at once. Subsequently, the new membrane potential will be transmitted to both pre- and post-synapse neighbours via network (the design before adjustments only sends the potential to the post-neurons).

4.4 Performance Evaluation

4.4.1 Model Configuration

To find an optimal design, first the limit is given on the total number of implementable physical cells $Total_{PhC}$ in FPGA, based on the available critical resources (4.6). The maximal number of clusters φ in the system depends on required accuracy. After dividing $Total_{PhC}$ by φ, the amount of physical cells per cluster (*PPC*) can be determined as

$$Total_{PhC} < \frac{\#Critical\ Resourse}{\#Resourse,\ \ PhC} \tag{4.6}$$

$$PPC = [\#Total_{PhC} \times \frac{1}{\varphi}], \text{ where } \varphi \leq 5 \tag{4.7}$$

The latency of physical cells cannot exceed the real-time constraint[1]. For each neuron, the latency cycle C_{neuron} consists of two parts: calculation cycle C_{cal} and communication cycle C_{com}

$$C_{neuron} = C_{cal} + C_{com} \tag{4.8}$$

The calculation cycle designates the time that neuron cell needs to generate the response, performs the calculations with the parameters and updates the results in the memory. The sending of the results to corresponding neighbours via the network is calculated in the communication cycles. The latency of the physical cell C_{PhC} is

$$C_{PhC} = TSF \times C_{cal} + C_{com} \tag{4.9}$$

[1]The 'real-time' constraint is the maximum number of cell states that can be computed within the model (with a step time of 50 μs given by [22]).

where TSF is the time-sharing factor for each physical cell. To improve the efficiency of system, C_{PhC} aims to be close to the real-time constraint, and consequently, a larger number of the neuron cells can compute the responses within a given system period T_{system} (4.11). From (4.9), an upper bound of the time-share factor can be calculated by considering the latency of *PhC*, the number of calculation cycles and the communication cycles

$$C_{system} = \frac{T_{system}}{CLK_{period}} \tag{4.10}$$

$$C_{PhC} \leq C_{real-time} \leq C_{system}, \quad where \quad C_{real-time} = 50\mu s \tag{4.11}$$

$$TSF \leq \left\lceil \frac{C_{PhC} - C_{com}}{C_{cal}} \right\rceil \tag{4.12}$$

As a result, the total number or neurons implemented in the system is derived by

$$Total_{neurons} = \varphi \times PCC \times TSF \tag{4.13}$$

The buffer depth is defined based on the layer where the corresponding router is placed in the tree network. Taking into consideration the number of physical cells per cluster and time-share factor, the maximal amount of packets in each cluster is

$$Num_{PhC,cluster} = N \times PPC \times TSF \tag{4.14}$$

where N is the maximum number of the connections to the neighbours. By adding the number of clusters in the design, the low bound on the buffer depth is given as

$$Depth_{downstream,\ ith} \geq 2^i \times N \times PPC \times TSF + Num_{other} \tag{4.15}$$

$$Depth_{upstream,\ ith} \geq 2^{i+1} \times N \times PPC \times TSF \tag{4.16}$$

4.4.2 Experimental Results

All simulations are completed with cycle-accurate SystemC, including all computation and communication latencies, both on- or off-chip. To simulate the system behaviour, three different connection schemes are utilized: all-to-all connections (all cells are connected to all other cells), normal-distributed distance-based connections (probability-based connections weighted by the distance between cells) and neighbour-based connections (connects every cell to every neighbouring cells, thus resulting in eight connections per cell).

The configuration file contains all the relevant parameters of the system and can be easily modified allowing exploration of different fan-out values, different cell communication schemes, etc. After the design is configured with the chosen accuracy (32/64-bit), it is synthesized through the Vivado HLS tool to generate VHDL code and test bench files. To assess the proposed design, the synthesizable VHDL code is compiled with ModelSim, and the simulated axon voltages are compared to the reference C model. Since the number of computational cycles is fixed (within a physical cell cluster) for a given topology, the hardware designs closest to real-time timing constraint are scaled up by increasing the number of physical cell clusters in the tree network. However, without routing tables in the tree network, all resulting potentials are sent in an all-to-all-type fashion.

The neuron spiking properties are governed by the specific parameter sets: these properties play well-defined role in defining explicit brain functions, e.g. the cortical neurons with tonic bursting contribute to the gamma-frequency oscillations in the brain. Figures 4.4 and 4.5 illustrate several types of neuron behaviour in the simulated system. Similar patterns are found with biological test. Most neurons are quiescent, but can fire spikes when stimulated. Figure 4.4(a) shows the typical firing of the integrate-and-fire model. When the pulses of the current are injected at the input, the neurons fire a train of spikes, and the process is called tonic spiking (Figure 4.4(b)). If such neurons fire continuously, it indicates that persistent input is offered to the neurons. A specific neuron could fire only a single spike at the onset of the input, and could subsequently stay quiescent, i.e. a response called phasic spiking, as illustrated in Figure 4.4(c). Specific neurons fire periodic bursts of spikes when stimulated, as shown in Figure 4.5(a).

Similar to the phasic spiking, the neurons can show phasic bursting behaviour, as in Figure 4.5(b), which is needed to transmit saliency of the input, to overcome the synaptic transmission failure and reduce neuronal noise, or can be used for selective communication between neurons. Intrinsically bursting excitatory neurons depicted in Figure 4.5(c) can exhibit a mixed type of spiking activity.

Table 4.1 lists hardware utilization of different spiking neuron models in the system. The system is implemented on the Virtex 7-XC7VX55 FPGA device. The hardware utilization is sub-divided into four different resource types, including look-up tables (LUT), flip-flops (FF), digital signal processors (DSP) and block memories (BRAM); smaller components like the synchronization circuits are omitted for clarity. FPGA resources can accommodate 1188 Hodgkin–Huxley-type neuron cells, and approximately

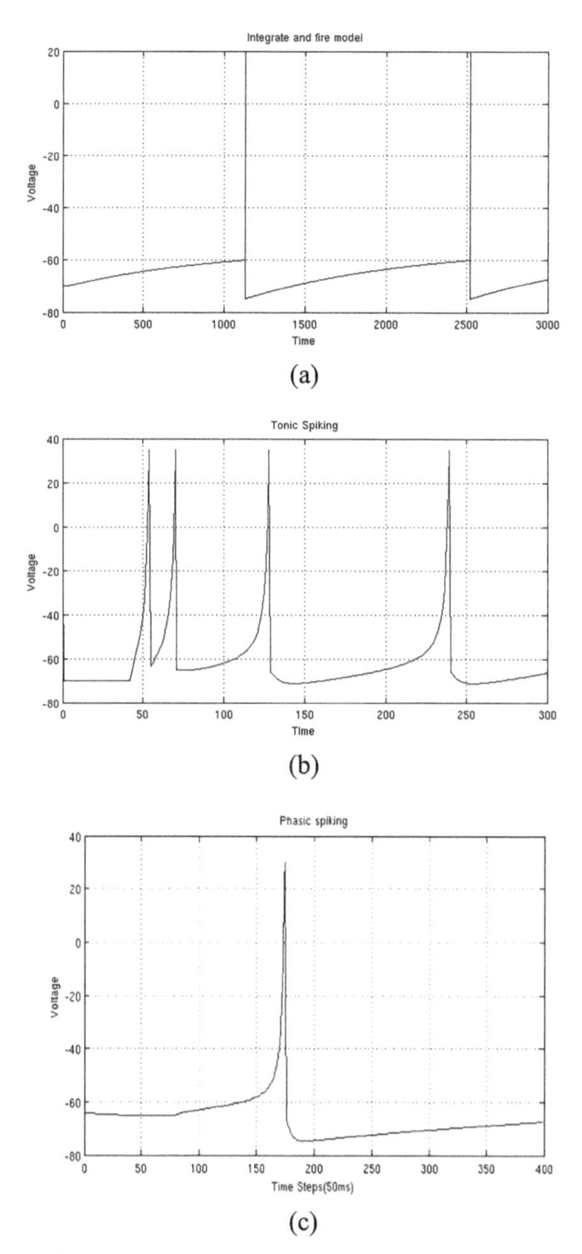

Figure 4.4 SystemC simulation of the neuro-computational properties of spiking neurons: (a) the integrate-and-fire model, (b) tonic spiking in the Izhikevich model and (c) phasic spiking in the Izhikevich model.

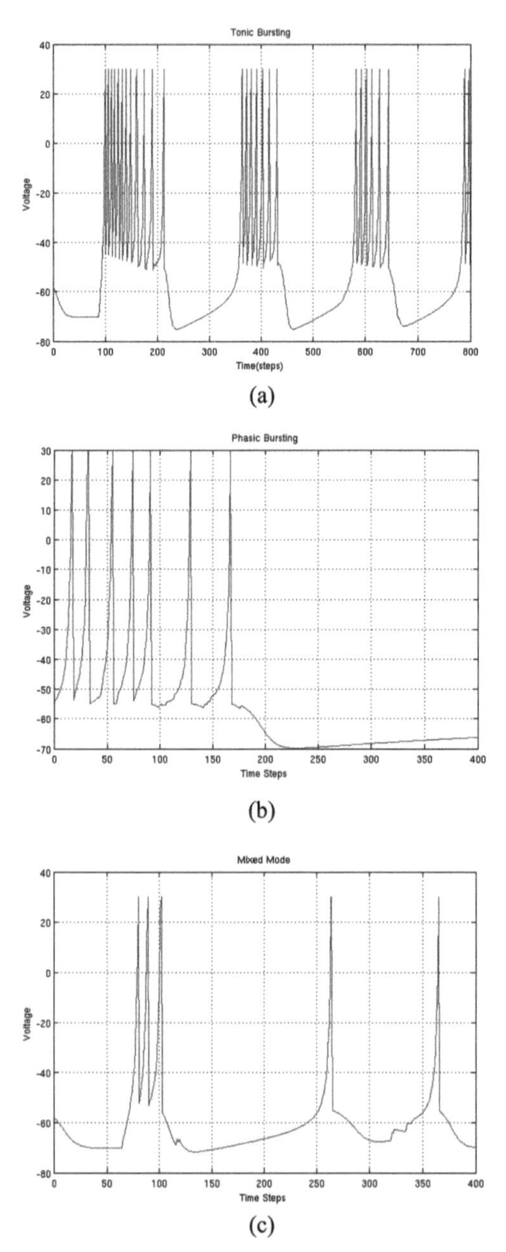

Figure 4.5 SystemC simulation of the neuro-computational properties of spiking neurons in the Izhikevich model: (a) tonic bursting, (b) phasic bursting and (c) mixed type of spiking activity.

Table 4.1 Hardware utilization of the most important components of the system on a Xilinx Virtex 7 XC7VX550 FPGA board

Model	Cluster	PhC	TSF	BRAM%	DSP%	FF%	LUT%	Neurons
Hodgkin–Huxley [7]	NA	8	12	78	57	27	83	96
Hodgkin–Huxley [4]	18	2	33	23.6	35	27.5	90	1188
Izhikevich	5	8	70	38	22	25	89	2800
Integrate and fire	5	8	75	23	20	16	54	3000

Routers sizes are generated by synthesizing a SystemC model of a router using Vivado HLS 2013.4.

2800 and 3000 Izhikevich and integrate-and–fire-type neural cells, respectively. The minimum simulation interval to achieve a realistic representation of the neuron-cell behaviour is determined as in [22]. Consequently, since each neuron cell can be re-used multiple times within the real-time boundary (and to benefit from the high level of parallelism and performance of the FPGA), the neuron cells are time-multiplexed. All results reported are for real-time simulations with double floating-point precision for the most biologically accurate representation of a neuron cell behaviour.

4.5 Conclusions

Real-time reconfigurable learning neuron networks are not only limited by run-time configurability, the re-synthesis of the system and the interconnection between the neurons, but mainly by the amount of neurons that can be placed on the chip. In this chapter, we implemented several models of the spiking neurons with axon conduction delays and spike timing-dependent plasticity (STDP) in a real-time data-flow learning network. The system implemented on the Virtex 7-XC7VX55 device can accommodate 1188 Hodgkin–Huxley-type neuron cells, and approximately 2800 and 3000 Izhikevich and integrate-and-fire-type cells, respectively. A tree-based communication bus is utilized, since it offers run-time configurability, e.g. the user-enabled configuration of the connectivity between cell and the calculation parameters adaptability. Consequently, the system does not need to be re-synthesized just to experiment with a different connectivity between cells. The cells are grouped around a shared memory in clusters to allow instantaneous communication.

References

[1] Abbott, L.F. "Lapique's introduction of the integrate-and-fire model neuron". In: Brain Research Bulletin 50, pp. 303–304, 1999.
[2] Brain Structure. url: https://psychbrains.wordpress.com/2014/10/07/brain-structure/.
[3] C. Morris and H. Lecar. "Voltage oscillations in the barnacle giant muscle fiber". In: Biophys. J 35, pp. 193–213, 1981.
[4] G.J. Christiaanse. "A Real-Time Hybrid Neuron Network for Highly Parallel Cognitive Systems". In: MSc Thesis, Delft, The Netherlands: Delft University of Technology (2016).
[5] O. College. Anatomy and Physiology. 2013. url: Http://cnx.org/content/%20col11496/latest/.
[6] M. van Eijk, et al. "Modeling of Olivocerebellar Neurons using SystemC and High Level Synthesis". In: IEEE, Biomedical Circuits and Systems Conference (2014).
[7] G. Smaragdos, et al. "FPGA-based Biophysically-meaningful Modeling of Olivocerebellar Neurons". In: Proceedings of the 2014 ACM/SIGDA International Symposium on Fieldprogrammable Gate Arrays, pp. 89–98, 2014.
[8] H. Gray. "Anatomy of the human body". In: Lea and Febiger (1918).
[9] GTKwave 3.3 Wave Analyzer User's Guide. url: http://gtkwave .sourceforge.net/gtkwave.pdf.
[10] H. R. Wilson. "Simplified dynamics of human and mammalian neocortical neurons". In: J. Theor. Biol 200, pp. 375–388, 1999.
[11] M. Tyrrell H. Shayani P. Bentley. "A cellular structure for online routing of FPGAs". In: Evolvable Systems: From Biology to Hardware, Springer Berlin-Heidelbarg, pp. 273–284, 2008.
[12] Hiroki Matsutani, Michihiro Koibuchi, Hideharu Amano. Tightly-coupled Multi-layer Topologies for 3-D NoCs.
[13] A. L. Hodgkin and A. F. Huxley. "A quantitative description of membrane current and its application to conduction and excitation in nerve". In: The Journal of Physiology, pp. 500–544, 1952.
[14] "IEEE Std 1666 – 2005 IEEE Standard SystemC Language Reference Manual". In: (2006).
[15] J. Hofmann. "Multi-chip dataflow architecture for massive scale biophysically accurate neuron simulation". In: MSc Thesis, Delft, The Netherlands: Delft University of Technology (2014).

[16] J. R. De Gruijl, P. Bazzigaluppi, M. T. G. de Jeu, and C. I. De Zeeuw. "Climbing fiber burst size and olivary sub-threshold oscillations in a network setting". In: PLoS Comput Biol 8.12 (2012).
[17] K. Cheung, S.R. Schultz, W. Luk. "A large-scale spiking neural network accelerator for FPGA systems". In: International Conference on Artificial Neural Networks and Machine Learning, pp. 113–120, 2012.
[18] Karsten Einwich et al. "Introduction to the SystemC AMS DRAFT standard". In: SoCC. IEEE (2009), p. 446. url: http://dblp.uni-trier.de/db/conf/%20socc/socc2009.html#EinwichGBV09.
[19] Kit Cheung, Simon R. Schultz, Wayne Luk. "A Large-scale Spiking Neural Network Accelerator for FPGA Systems". In: Conference on Artificial Neural Networks and Machine Learning (2012).
[20] Baptiste Lepilleur. JSONCpp. url: https://github.com/open-source-parsers/jsoncpp.
[21] Levy WB, Steward O. "Temporal contiguity requirements for long-term associative potentiation/depression in the hippocampus". In: Neuroscience 8.4, pp. 791–797, 1983.
[22] M. van Eijk, C. Galuzzi, A. Zjajo, G. Smaragdos, C. Strydis, R. van Leuken. "ESL design of customizable real-time neuron networks". In: IEEE International Biomedical Circuits and Systems Conference, pp. 671–674, 2014.
[23] Wolfgang Maass. "Networks of spiking neurons: The third generation of neural network models". In: Neural Network 10.9, pp. 1659–1671, 1997.
[24] MAK Shrimawale, MA Gaikwad. Performance Analysis of Odd-Even Routing Algorithm of Network-on-chip Architecture for 2D mesh topology under Bursty Communication Traffic.
[25] Marcel Beuler, Aubin Tchaptchet, Werner Bonath, Svetlana Postnova, Hans Albert Braun. "Real-Time Simulations of Synchronization in a Conductance-Based Neuronal Network with a Digital FPGA Hardware-Core". In: Artificial Neural Networks and Machine Learning, pp. 97–104, 2012.
[26] J.V. Alvarez-Icaza Merolla P.A. Arthur. "A million spiking-neuron integrated circuit with a scalable communication network and interface". In: Science 345 (2014).
[27] Engene M. Izhikevich. Which Model to Use for Cortical Spiking Neurons? url: http://www.izhikevich.org/publications/whichmod.htm.

[28] Engene M. Izhikevich. “Simple Model of Spiking Neurons”. In: IEEE Transactions on Neural Networks, vol. 14, no. 6, (2003).

[29] Engene M. Izhikevich. “Polychronization: Computation with Spikes”. In: Neural Computation 18, pp. 245–282, 2006.

[30] Emin Orhan. The Leaky Integrate-and-Fire Neuron Model. 2012. url: http://www.cns.nyu.edu/∼eorhan/notes/lif-neuron.pdf.

[31] P. Parandkar, J.K. Dalal, S. Katiyal. “Performance Comparison of XY, OE and DY Ad Routing Algorithm by Load Variation Analysis of 2-Dimensional Mesh Topology Based Network-on-Chip”. In: Bvicams International Journal of Information Technology (2012).

[32] Peter J. Bentley, Andy M. Tyrrell. “Hardware Implementation of a Bio-plausible Neuron Model for Evolution and Growth of Spiking Neural Networks on FPGA”. In: Adaptive Hardware and Systems (2008).

[33] Q. Yu, R. Yan, H. Tang, K.C. Tan, H. Li. “A spiking neural network system for robust sequence recognition”. In: IEEE Transactions on Neural Networks and Learning Systems 19.4, pp. 621–635, 2016.

[34] R. M. Rose and J. L. Hindmarsh. “The assembly of ionic currents in a thalamic neuron. I The three-dimensional model”. In: Proc. R. Soc. Lond.B 237, pp. 267–288, 1989.

[35] Resve Saleh, Michael Jones, Andre Ivanov, Partha Pratim Pande, Cristian Grecu. “Performance Evaluation and Design Trade-O_s for Network-on-Chip Interconnect Architectures”. In: IEEE Transactions on Computers 54, pp. 1025–1040, 2005.

[36] Reza Sabbaghi-Nadooshan, Mehdi Modarressi, Hamid Sarbazi-Azad. “The 2D DBM: An attractive alternative to the simple 2D mesh topology for on-chip networks”. In: IEEE International Conference on Computer Design (2008).

[37] Guido van Rossum. Python Programming Language. url: https://www.python.org/.

[38] Rumsey CC, Abbott LF. “Equalization of synaptic efficacy by activity- and timing-dependent synaptic plasticity”. In: Neurophysiology 91.5, pp. 2273–2280, 2004.

[39] H. Adeli S. Ghosh-Dastidar. “Spiking neural networks”. In: International Journal of Neural Systems 19.4, pp. 295–308, 2009.

[40] Sebastien Le Beux, Ian O’Connor. “Reduction methods for adapting optical network on chip typologies to 3D architectures”. In: Microprocessors and Microsystems 37, pp. 87–98, 2013.

[41] The Integrate-and-Fire Neuron Model. url: http://neuroscience.ucdavis.edu/goldman/Tutorials_files/Integrate%26Fire.pdf.
[42] W. Gerstner, W.M. Kistler. "Spiking neuron models: single neurons, populations, plasticity". In: Cambridge University Press (2002).
[43] J. P. Welsh and R. Llinas. "Some organizing principles for the control of movement based on olivocerebellar physiology". In: Progress in Brain Research, pp. 449–461, 1997.

5

Energy-Efficient Multipath Ring Network for Heterogeneous Clustered Neuronal Arrays

Andrei Ardelean

Delft University of Technology, Delft, The Netherlands

Simulating large spiking neural networks (SNN) with a high level of realism in a field-programmable gate array (FPGA) requires efficient network architectures that satisfy both resource and interconnect constraints, as well as changes in traffic patterns due to learning processes. In this chapter, based on a clustered SNN simulator concept, an energy-efficient multipath ring network topology for the neuron-to-neuron communication is proposed. The topology is compared in terms of its mathematical properties with other common network topology graphs after which the traffic distributions across the network are estimated. As a final characterization step, the energy-delay product of the multipath topology is estimated and compared with other low-power architectures. In addition, a simplified binary tree is suggested as a network layer for handling configuration and input/output data that use a custom channel protocol without the need for routing tables.

5.1 Introduction

Spiking neural networks (SNN) have become increasingly more attractive in scientific research recently due to their ability to closely mimic biological neural behavior such as the encoding of information in the timing of spikes and their amplitude and in spike train patterns and transfer rate [1]. However, such high level of realism comes at the cost of substantial computational effort and high-throughput demands. Multiple neuron models such as leaky integrate-and-fire [2], Izhikevich [3], and Hodgkin–Huxley [4] have been developed and implemented in hardware, with each providing different levels of realism. However, the need to create flexible simulating hardware capable of implementing at the same time multiple model types with large number of neurons persists. Because neural networks present a high level of parallelism in their behavior, field-programmable gate arrays (FPGA) prove to be suitable platforms for SNN simulation [5]. In addition, they also benefit from inherent flexibility that allows them to change simulated network topologies and neuron models with ease. However, the interconnect fabric in FPGA is limited in terms of availability, power, and delay [6], and therefore, network topologies have to be developed that efficiently use FPGA resources.

In addition to the aforementioned issues, neuron-to-neuron communication schemes are continuously changing during simulation time because of learning processes [7]. Connections between neurons that existed at the beginning of the simulation can disappear while other new ones can be formed. Assuming an intelligent placement of the neurons based on an initial communication scheme, the resulting changes can mitigate the effect of all the optimizations. These modifications in traffic can lead to inefficiencies in the overall system (such as inadmissible latency and loss of packets due to contention) if the simulating platform is not flexible enough to accommodate them. Solutions might include not only the use of dynamic routing algorithms or neuron placement reconfiguration, but also the design of efficient network architectures that can manage such changes due to their topological properties (small diameter, high degree of connectivity, etc.).

This chapter is organized as follows: Section 5.2 summarizes basic concepts related to the studied topic and state-of-the-art research in the field while Section 5.3 presents the analyzed neuron communication schemes and defines the proposed system structure as well as assumptions on which it is based. Section 5.4 focuses on the system architecture and its analysis and compares the multipath ring and two-dimensional torus network topologies, as well as presents the energy-delay product estimation for different network topologies. Section 5.5 concludes the chapter.

5.2 State-of-the-Art and Background Concepts

5.2.1 Neuron Models

The *neuron*, also called a *nerve cell*, is an electrically excitable cell that receives, processes, and then transmits information in the form of electrical and chemical signals. Neurons mainly consist of three parts: the *dendrites*, through which inputs from other neurons are received, the *soma* or the cell body, and the *axon*, which acts as the neuron output. A voltage gradient is maintained across the cell membrane due to varying ion concentrations in the intra- and extracellular media. If the voltage across the membrane changes by a significant amount, an electrochemical pulse known as *action potential* is generated and propagated along the axon activating the connections with other neurons. These connections between neurons are called *synapses*, with cells in the human brain being capable of reaching 10k connections with their neighbors. A series of interconnected neurons is referred to as a *neural network.*

A multitude of models have been proposed to describe the behavior of a neural cell. This chapter will be based on a system designed for the extended Hodgkin–Huxley model [11]. This model describes the *inferior olivary nucleus (ION)* as a multiple compartmental cell, with each compartment having a state potential updated every simulation cycle that depends on dendritic potentials and ion currents. The model structure is shown in Figure 5.1. It includes an extra compartment to model the axon hillock as well as the possibility to connect the dendritic compartment to other neighboring cells. Compared with other neuronal models, the Hodgkin–Huxley model is the most biophysically accurate (Figure 5.2) as it describes the membrane

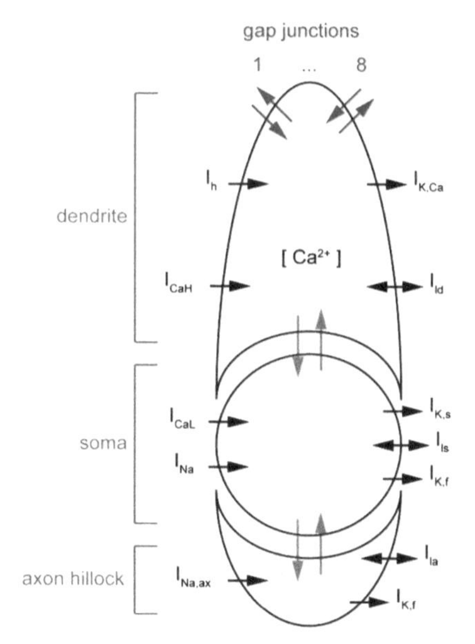

Figure 5.1 The three-compartment-based extended Hodgkin–Huxley neuron model [11].

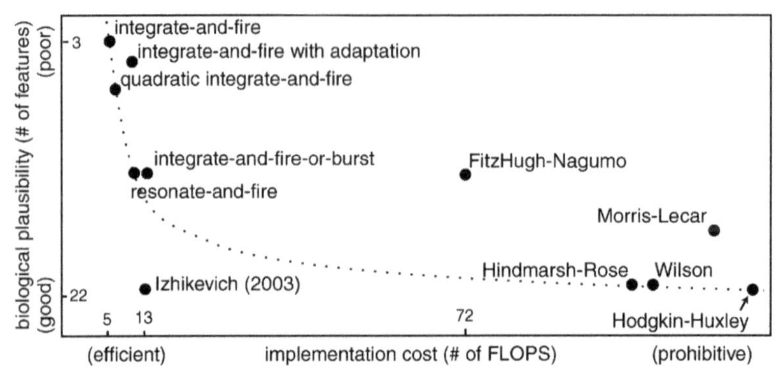

Figure 5.2 Comparison of different neuron models in terms of computational complexity and biophysical accuracy [12].

potential and activation of ion currents. It can also exhibit all behaviors of biological spiking neurons (tonic spiking, phasic spiking, bursting, etc.) by tuning the tens of parameters that it comprises [12]. However, this level of realism requires a large computational effort, and very few neural simulators implement it for cases where large number of neurons are desired.

5.2.2 Simulation Platforms

Several approaches have been suggested for simulating SNN with various degrees of biological realism on multiple platforms such as CPU and/or GPU, analog–digital hybrids, ASIC, and FPGA.

*SpiNNake*r [13] is a massively parallel computer that is targeted to accommodate up to 1 million ARM microprocessor cores with a communication architecture optimized for carrying large numbers of small data packets. The microprocessor cores are distributed throughout 57,000 nodes [14], each one containing 18 ARM cores, 128 MB of shared memory, 96 kB of local memory, a router, and general-purpose circuitry. Six communication links connect each node in a triangular fashion folded on the surface of a toroid [14] with a routing scheme based on routing tables with 1024 entries. The system is designed to simulate leaky integrate-and-fire or Izhikevich neuron model with the order of 1000 input synapses to each neuron, one ARM core being capable of modeling 100 neurons and theoretically reach 10 million connections [13].

Neurogrid [15] is a hybrid architecture that consists of Neurocores, a combination of 256 $\times$ 256 analog neuron cells with digital routing logic and memory. The communication network topology is a binary tree for which a version of up-down routing with a point-to-point and then branching phase was developed in order to guarantee that the network is deadlock-free [16]. The routing path is encoded in the packet header with each up or down and left and right turns represented by a single bit, depending on the phase in which the packet is at that moment.

TrueNorth [17] is a digital ASIC implementation of an SNN developed by IBM, consisting of 4096 neurosynaptic cores that each simulates 256 integrate-and-fire neurons with 256 configurable synapses. The cores are distributed in a two-dimensional mesh, where X-Y routing is employed to eliminate deadlocking. The result is an energy-efficient system that can be easily scaled to multi-chip configurations.

NepteronCore [18] is a digital neural core implemented on a Xilinx Virtex 4 FPGA that simulates a simplified version of the Hodgkin–Huxley model called the Huber–Braun model [19] that only focuses on clinically relevant measures. In total, 1600 neurons are simulated with a model step size of 100 μs [20]. However, the design is still at the stage of having only one core that takes up around 16% of FPGA resources.

PLAQIF (piecewise-linear approximation of quadratic integrate and fire) is a digital neuron model implemented by [21] on a Xilinx Virtex 5 FPGA. A network consisting of 161 neurons and 1610 synapses was simulated at a 1 ms simulation step. Due to a design decision to make the dendrite structure flexible during learning processes, FPGA resources are heavily used, with the design occupying 85% of the device.

In [22], an Izhikevich model-based SNN used for character recognition is implemented in FPGA. The design uses 25 processing elements (PE) to compute all the required calculations for a total of 9000 neurons. The design was implemented on a Xilinx Virtex II Pro FPGA and consumed 79% of available resources. The disadvantage of this architecture is that all computations are in fixed-point representation.

Brainbow [23] is a project aimed to simulate Izhikevich modeled neurons using a 1 ms time step. A computation core is built around a single multiplier in a Xilinx Virtex 4 FPGA that is then used to simulate a network of 117 neurons with only 3.3% resource usage. However, the limitations of this architecture arise from the memory usage and, as a result, future implementations would require external RAM memory modules.

Overall, the various implementations mentioned above present specific advantages and disadvantages that arise from the platform of choice. ARM- and GPU-based systems benefit from the flexibility of being able to change neuron models but lack the ability to efficiently map the highly parallel nature of SNN and, as a result, are also the slowest implementations. ASICs provide the base for the most energy-efficient systems that also exhibit the highest simulation speeds but are extremely rigid, as the neuron models are fixed. A good compromise between the two are FPGA-based platforms that are capable of efficiently executing parallel computations while at the same time allowing a reconfiguration of the networks as desired by the user. The limited number of available resources, however, restricts the network size, but this can be overcome by multi-chip solutions.

5.2.3 Communication Network Considerations

The communication network is tasked with delivering data packets between all the clusters in the system in an efficient manner. It consists of a number of routers interconnected with links, on top of which a routing protocol, encoded in the routers themselves, manages packet delivery. The links are composed of wires and can be categorized into two types: parallel and serial. The parallel links can be operated at low clock rates to reduce power consumption, but use large amounts of hardware resources, especially when data has 32 or 64 bit format as in this case. Contrastingly, serial links operate at high clock speeds and require serialization and de-serialization hardware, but consume less interconnect resources. The links ultimately define the performance and power consumption in the network and the main design objectives are to provide fast, reliable, and low-power interconnects between the nodes of the network [24].

The multipath ring characterization performed will not define the type of links but instead will make them more abstract by considering the link as an ideal medium for transmitting data from one end to another. The actual implementation type can be selected once the throughput is determined. However, when estimating delay and power consumption of various network topologies, the links will be given delay and energy metrics.

Routers are usually composed of a number of ports and a switch matrix. The ports can be further classified into network ports that are connected to links in the network and local ports that are dedicated to the IP core attached to the router, in this case the cluster. Apart from the physical interconnect infrastructure, additional circuitry is present, which handles the flow of data through the router based on a set of predefined rules called the *routing protocol/scheme*.

Performance of a network can be evaluated through a number of parameters:

Bandwidth – the maximum rate of data propagation measured in bits per second (bps). This parameter will not be used in this chapter because the physical network layer is not implemented and hence no protocols are defined.

Throughput – the maximum traffic accepted by the network measured in messages per clock cycle [25]. Throughout this chapter, throughput will be measured in messages per simulation step.

Latency – the time it takes for a message to be completely received by the target, counting from the beginning of transmission, measured in time units. This parameter will be used to rank different network graphs as a first characterization step of the multipath ring. It will not be directly calculated but will be qualitatively determined based on mathematical properties such as average path length and network diameter.

The overall structure of router connections is referred to as a *topology* and it can be *direct*, where each router is connected to an IP core (cluster) and *indirect* where some routers are only used to propagate messages through the network while others are connected to cores and can serve as message sources or destinations [24]. Both topology types are analyzed; however, the networks with the former type are preferred since they are more hardware-efficient. For the same reason, *regular networks*, where every router has the same number of ports, are the only ones analyzed.

Table 5.1 contains a collection of network topologies along with their characteristics. Included among them is also the multipath ring and two-dimensional torus that will be further described in Sections 5.4.2.1 and 5.4.2.2.

The routing protocol is the set of rules based on which a router output port is selected for a specific message that needs to pass through. Usually, this implies the use of routing tables and address information present in the packet header. Two routing categories can be identified: (1) *static routing,* in which the path a message follows between two nodes is always the same and is determined before runtime and (2) *dynamic routing*, in which the routing tables can be modified during runtime depending on the current status of the network (if, for example, the original path is congested). The state-of-the-art implementations presented in the previous section employ three network topologies: binary tree, mesh, and torus with both static (*Neurogrid* and *TrueNorth*) and adaptive (*SpiNNaker*) routing. The use of adaptive routing allows the system to more efficiently utilize available resources at run time but these optimizations are limited by the network structure. Also, highly

Table 5.1 Common network topologies

Type	Description	Advantages	Disadvantages
Ring	Every router is connected to two others and a circular path is formed	– Simple routing scheme – Simple HW implementation	– Large diameter – Deadlock possibility
Multipath ring	Similar to the ring but with an extra set of connections between routers that are further apart	– Small diameter – Flexible – Simple HW implementation	– Larger average path length for big networks
Binary tree	There is only one connection between any two connected routers, forming a parent–child hierarchy. Each node has two children	– Small diameter – Recursive structure	– Long interconnects between routers at root increase delay and power consumption
Mesh	Routers are arranged in an orthogonal array and each connected to its immediate neighbors	– Efficient layout implementations – Simple routing	– Large diameter
Two-dimensional torus	Similar to the mesh but the routers on the edges have wrap-around connections	– Small diameter – Simple routing	– Longer delay in wrap-around connections

adaptive algorithms are expensive in terms of virtual channels and, as a result, have large-area overhead. Reducing the number of virtual channels would reduce the efficiency of the approach [26].

An interesting alternative would be to change the network topology as changes in traffic patterns take place, such as the addition or removal of routers in the network [27] or changing link positions [28]. Unfortunately, these approaches consume a lot of hardware resources and represent an overcomplication of the network topology. If, however, a flexible architecture that requires only a small number of reconfigurable links and routers to manifest large changes in traffic distribution were developed, these techniques could be implemented with less hardware overhead. Investigations in terms of traffic redistribution in the multipath ring were carried out that indicate a flexible architecture that can represent a good candidate for these kinds of applications.

5.3 Neural Network Communication Schemes and System Structure

A set of preliminary analysis steps are necessary in order to define the design space of the system. There are two factors that must be taken into account:

The neural network communication scheme is different from the physical network implementation due to hardware constraints. As noted in Section 5.2.1, neurons can reach up to 10 k synapses each, which makes it difficult to translate them into hardware representations, let alone networks that contain large numbers of neurons, since the result would be slow, or would require a very large amount of resources. In addition, Section 2.3 shows a number of optimizations in the form of clusters that are used to boost system performance, structures that have no biological equivalent.

The neuron communication scheme will dynamically change during run time in an uncontrolled fashion. Throughout the simulation process, multiple parameters such as concentration levels used in the model and cell-to-cell connections change as a result of user intervention or learning processes undergone by the network. These changes have a significant impact on the traffic behavior, and if the physical network cannot handle them, they can lead to deadlocking, loss of packets, and inability to maintain real-time operation.

For simplification purposes, the neural network communication scheme is characterized by a set of simple metrics presented in the following section that are then used as guidelines for designing the overall system structure.

5.3.1 Physical System Structure

The neuron-to-neuron communication scheme is commonly represented in the form of a graph, in which a neuron is represented by a node (vertex) and a connection is a directed edge (Figure 5.3). The effect a neuron has on the network is linked to the number of connections it has with other neurons, a number equal to the degree d_i of the node, where i is the neuron (node) index. Typically, neurons can be classified in separate categories (or groups called *layers*) according to their function [29]. Neurons that receive stimuli from sources not in the network are called *input neurons* and those that send responses outside of the network are appropriately called *output neurons*. Between them, *hidden neurons* comprise the neural network bulk and do not interact with the outside.

Arguably the most common [30], the *feed forward* scheme (Figure 5.4(a)) is characterized by a signal path that goes from the input to the output layer

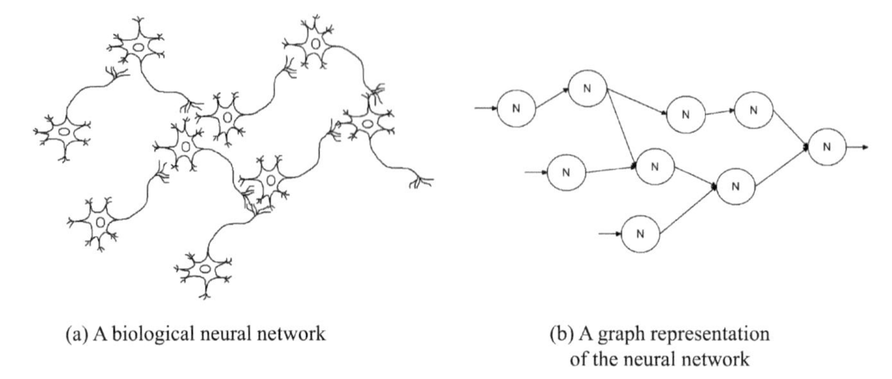

(a) A biological neural network

(b) A graph representation of the neural network

Figure 5.3 Representation of a biological network as a directed graph.

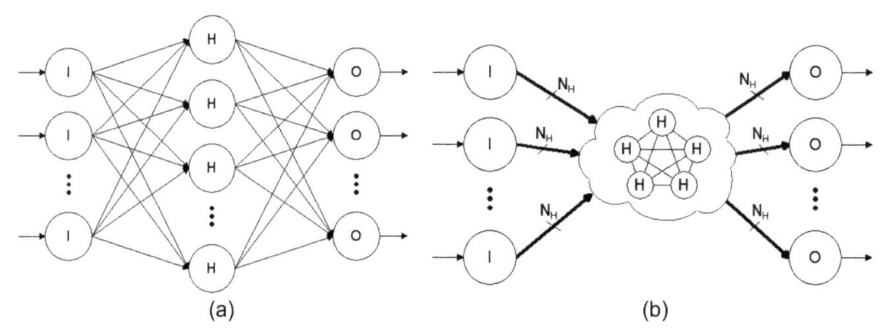

Figure 5.4 (a) Feed-forward communication scheme, (b) complete hidden layer communication scheme.

through a number of hidden layers, without forming any loops. In the case of multiple hidden layers, the same rules apply, with hidden (H) neurons from one layer taking their inputs from all the neurons in the preceding layer, and transmitting their outputs to all the neurons in the following layer.

The feed-forward communication scheme is incapable of mimicking biological processes such as memory because the outputs are only dependent on the current inputs. In recurrent neural networks, neurons are allowed to have connections that form a cycle, by connecting to either their own inputs or the inputs of neurons in the previous layer. The most prominent example of such a topology is the Hopfield network [31], in which all the neurons are connected with each other. All recurrent neural networks can be included in the *complete hidden layer* scheme, which, as the name suggests, has a hidden layer composed of neurons that are all connected to each other with bidirectional links as well as having their inputs and outputs connected to the I and O neurons, respectively (Figure 5.4(b)). Apart from the excitatory neuron behavior modeled by feed forward and complete hidden layer communication schemes, research in biology has also detected inhibitory behavior between neurons [32]. In the *layered full lateral inhibition* scheme, the neurons in the hidden layers are split into two groups: those with normal behavior and those with inhibitory functions. For each normal neuron (H) in the hidden layer, there exists an inhibitory one (H′) that takes its input from the former and outputs to all the other non-inhibitory neurons in the layer (Figure 5.5(a)). The *layered neighbor lateral inhibition* is similar to the full inhibition one described previously, with the change that an inhibitory neuron's outputs only connect to the immediate neighbors of its paired normal behavior neuron (Figure 5.5(b)).

Figure 5.6 illustrates the average degrees (a measure of how tightly the neurons are interconnected and as a consequence the network hardware requirements) of the four communication schemes as the number of neurons in the hidden layer increases form 10% to 90% of the total number of neurons N. The vertical axis is in percentages of N in order to normalize the results.

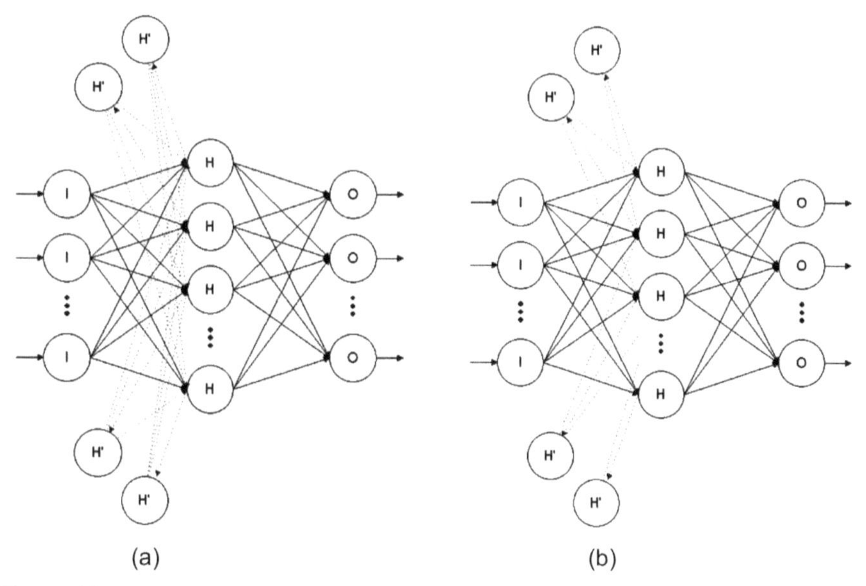

Figure 5.5 (a) Layered full lateral inhibition communication scheme, (b) Layered neighbor lateral inhibition communication scheme.

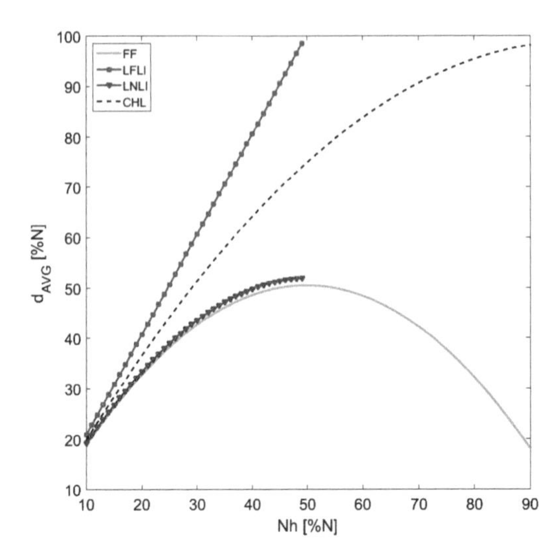

Figure 5.6 Average degree as a function of the number of hidden neurons in the four communication schemes.

Apart from the neuron-to-neuron communications, data traffic through the system also contains configuration packets, input stimuli, and output data, all of which have specific characteristics:

Configuration data are sent during setup or between simulation steps. It consists of initialization or update values of neuron model parameters with each targeted cluster receiving up to 29 double precision floating-point packets per neuron [8]. They represent ion concentrations and currents along with axon and soma potentials.

Input (stimulus) data are sent between simulation steps; they originate form a system controller (user) and are targeted toward input neurons. Analogous to the configuration data, they consist of multiple floating-point packets sent to each cluster. They represent voltage samples of various impulses or trains of impulses.

Output data are sent during run time toward the user and it originates from specific neurons/clusters previously selected by her/him. It consists of various parameters' values and has packets of varying size.

Because these behaviors have little in common with the neuron-to-neuron communication and are also deterministic in nature, it is worthwhile to create a physical layer dedicated to them that will have a more suited topology that increases the overall system efficiency.

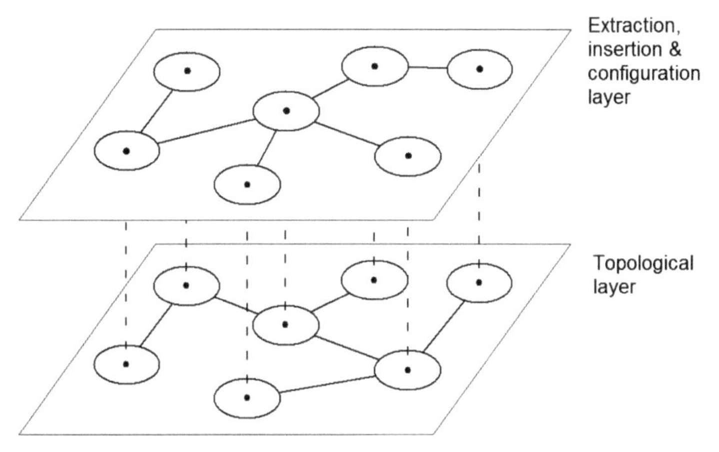

Figure 5.7 Proposed system structure.

The proposed system topology, shown in Figure 5.7, consists of two separate layers: (1) *extraction, insertion, and configuration layer*, which handles the output/input data transfers and (2) the *topological layer*, which handles the neuron-to-neuron communication.

These two layers are physically independent and only connect through the routers dedicated to each cluster. They have distinct topologies and follow different routing protocols, which will be described in the following chapter.

5.3.2 Extraction, Insertion, and Configuration Layer

This layer is dedicated to transferring configuration and input (stimulus) data to each cluster (targeting each neuron) and extracting the desired parameters and transmitting them to the user. The particularities of such type of data traffic are exploited so as to create an efficient network topology.

The downstream traffic (toward the clusters) in this layer has the particular trait of consisting of a large number of packets sent to the same destination. In the majority of cases, routing decisions are made based on consulting tables after the destination address of the packet is read from its header. The presence of this header and the time it takes to access the routing tables represent a protocol overhead. Even if a more complex scheme of reading the address header once and then delivering the following packets up to a special ending one was used, routing tables would still be required. For this reason, a *channel* protocol is suggested as an efficient replacement that eliminates the use of routing tables and reduces the overhead.

The upstream traffic (from the clusters) presents no remarkable characteristic traits, but it does not require a large bandwidth and is predictive. Upstream routing can be simplified to the point where no routing tables are used at all because there is only one possible destination, the user. In this case, routers are only tasked with handling arbitration in case of conflicts between data packets.

The physical implementation for the extraction, insertion, and configuration layer is a binary tree, with each leaf being a cluster and the root an off-chip connection point (Figure 5.8). Each router has one upstream and two downstream ports and a 1-bit configuration signal that is used in the channel routing.

With the $S/\overline{T}$ (*Setup/Transmit*) signal set high, each router reads its corresponding bit (related to the layer on which the router is positioned) from the setup message and redirects the package to either the left or right output port (Figure 5.9(a)). On the falling edge of $S/\overline{T}$, the setup is latched into each router and a direct channel is now created between the root and the destination cluster (Figure 5.9(b)) that will remain in place until $S/\overline{T}$ is reasserted. In this situation, packets sent from the root are automatically directed to the destination cluster.

5.3.3 Topological Layer

Selecting an appropriate physical network for the topological layer represents the main challenge of this design. The traffic is pattern-less and continually

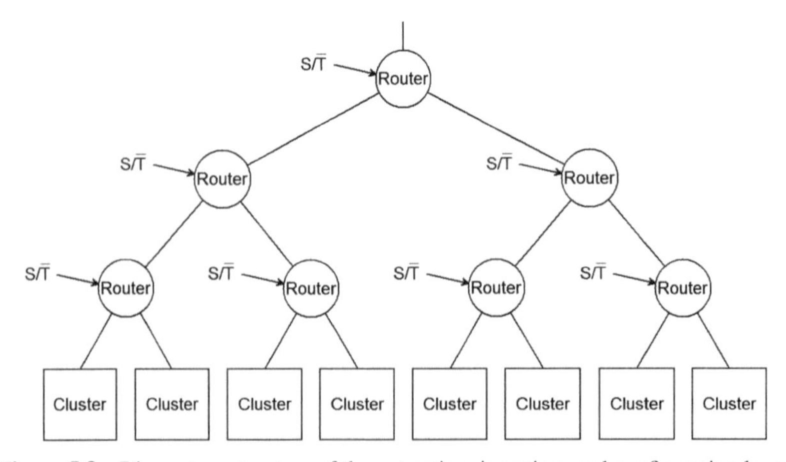

Figure 5.8 Binary tree structure of the extraction, insertion, and configuration layer.

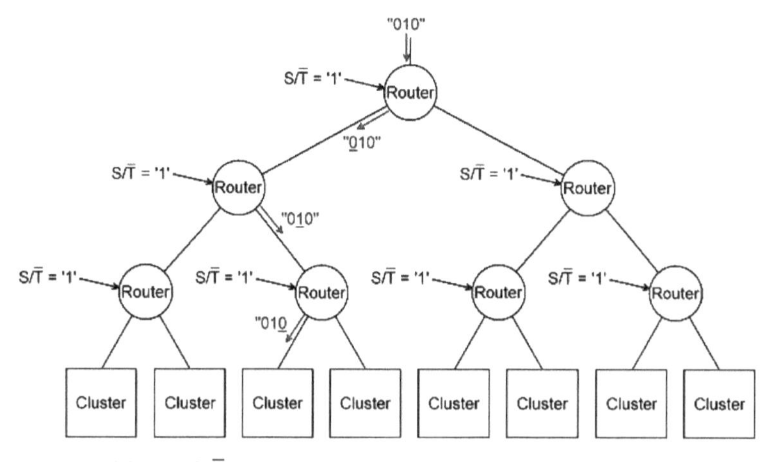

(a) With $S/\overline{T}$ asserted, each router reads its corresponding bit from the setup packet and routes it appropriately

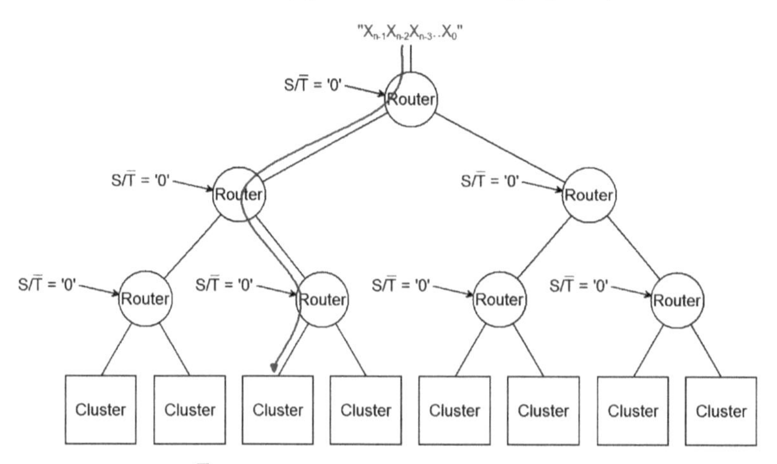

(b) With $S/\overline{T}$ set low, the previous setup is latched and a channel is maintained between the root and the targeted cluster

Figure 5.9 Steps in the routing scheme.

changing as the network goes through the learning process and with power and delay as optimization targets, the chosen topology has to have high connectivity with a relatively simple hardware implementation. Two network topologies are considered as possible solutions, the *Multipath ring* and *Two-dimensional torus*, with their descriptions and analyses in the following subsections.

5.3.3.1 Multipath ring routing scheme

The multipath ring (MPR) topology is an example of a regular graph with all node degrees (router ports) equal to 5 (one port dedicated to an associated cluster and 4 to the network) so as to maintain physical implementation complexity low. As with other graphs from the same family, MPR has a high clustering coefficient, but a relatively large path length for large networks [34]. However, assuming that data locality is exploited during neuron placement, the probability of a message needing to be sent to a cluster that is far away is small and the longer paths of the MPR will be rarely used. The expected number of clusters will not exceed 25 (due to the limitation of the current FPGA technology), which makes the disadvantages of this topology limited. The MPR links can be classified into two categories: *exterior* links, the ones connecting two nodes that are physically next to each other, and *interior* links, connecting nodes that are farther apart. These former links can have varying lengths and, as a consequence, can *skip* over a number of nodes and modify the average path length. Figure 5.10 shows an example of a multipath ring network of size 13 with skips of 1, 2, and 3. A small reduction in path length, from 2 to 1.83, is seen when increasing the skip, which comes at the cost of increasing traffic through the external links.

In order to estimate the traffic through the multipath ring topology, a routing scheme needs to be considered. For simplicity purposes, we assume static routing (the path between two clusters is known before run time and is always the same), with the following set of rules: If the message arrives from another router and the destination is this router's associated cluster, then send it to the cluster controller. If the message arrives from another router and the index distance to the destination cluster is less than or equal to the skip value, then send the message through the exterior link, otherwise through

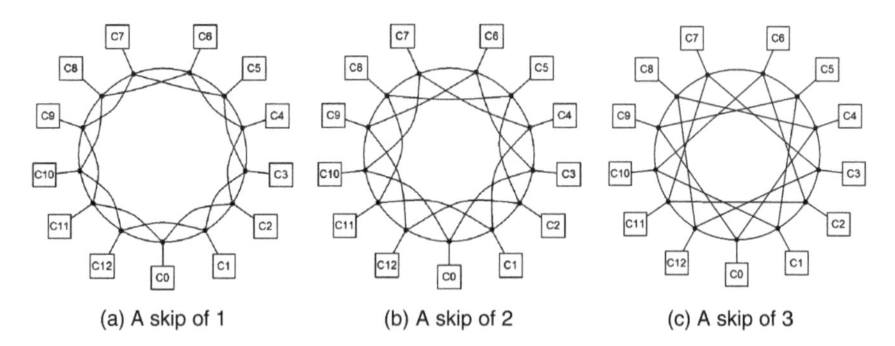

Figure 5.10 Multipath ring topologies of size 13 with various skip values.

the interior link. If the message originates from the cluster associated to this router and the distance to the destination cluster index is less than or equal to half the number of clusters, then send it counterclockwise; otherwise, send it clockwise. In both cases, pick an exterior or interior link like that in 2.

In essence, a router will prioritize sending a packet in the smallest number of hops and if the message originates from the associated cluster, then it will balance the network as much as possible by taking advantage of MPR symmetry. The *index distance* is equal to the Manhattan distance between two routers when only exterior links are considered. Figure 5.11(a) shows the resulting routing path for every message originating from cluster zero towards the rest. The symmetric nature of this scheme makes the MPR most efficient for an odd number of clusters (the clockwise and counterclockwise branches are equal) but it can also accommodate even numbers by extending one of the branches to the extra cluster. Deadlocking is avoided by not allowing packets to go from an exterior link to an interior one, similar to turn restriction routing in the mesh topology.

The two-dimensional (2D) torus topology is formed by arranging all the routers in a rectangular lattice with each one connected to its immediate neighbors and with the connections at the edges wrapping around. Figure 5.11(b) shows a 2D torus with 12 clusters, where each router is limited to 5 ports like in the MPR. The benefits of having such a topology come from the layout efficiency of the mesh and the existence of simple deadlock-free routing algorithms (X–Y, west first, north last, etc.) combined with the small

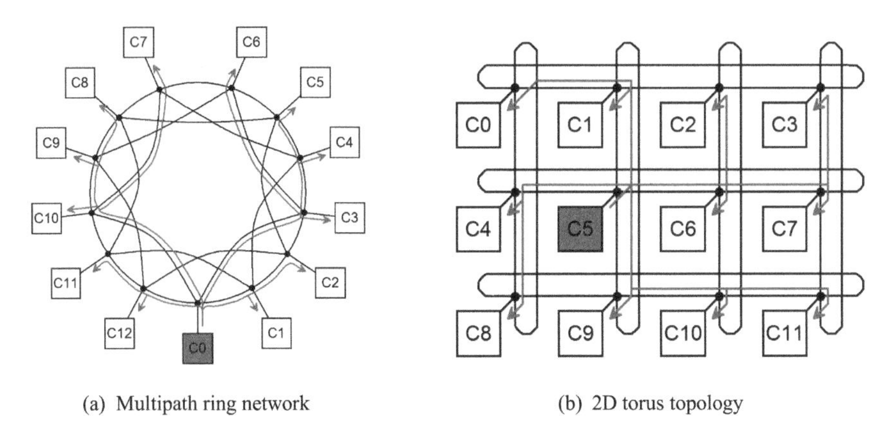

(a) Multipath ring network (b) 2D torus topology

Figure 5.11 A routing scheme for the multipath ring and 2D torus topology, respectively. The red arrows indicate the path taken by a message originating from cluster zero to every possible destination cluster.

network diameter given by the wrap-around connections. When a message originates from a cluster, it is sent on one of the four network ports based on the quadrant in which the destination is located, after which XY routing rules are followed. Figure 5.11(b) shows the resulting paths for messages originating from cluster five (chosen for illustration purposes due to its center position in the figure).

5.3.3.2 Traffic model

Due to the large computational effort required by the analysis of neural networks consisting of several thousands of biophysically accurate neuron cells, a set of assumptions are made that simplify the procedure: (i) a neuron cell contributes to the network traffic with an output rate, which depends on the model complexity and is known at the beginning of the simulation, (ii) all data packets are of same size, and (iii) we assume there is no bandwidth limit in the network links, as this is one of the metrics that will result from the analysis.

Considering these assumptions, we can compute an average output rate λ of all neurons in the network, which may be expressed in *packets/step* and use it as a measurement unit for the throughput estimates. The traffic estimation is based on three parameters: λ_i is the number of distinct outgoing messages from cluster *i*, $\mu_{i,j}$ is the number of messages from cluster *i* that are destined for cluster *j*, and η is the average percentage of outgoing messages common to all destination clusters.

Message refers to a data packet of arbitrary size that is sent from a cluster and contains the relevant neuron-to-neuron communication data. These parameters can be computed once all the neurons are placed in their respective clusters based on the communication scheme between them. The output metrics of interest are: ω_i is the throughput value associated with the counterclockwise direction on exterior links for MPR and east direction for 2D Torus, $\omega_i{}'$ is the throughput value associated with the clockwise direction on exterior links for MPR and west direction for 2D Torus, θ_i is the throughput value associated with the counterclockwise direction on interior links for MPR and north direction for 2D Torus, and θ_i' is the throughput value associated with the clockwise direction on interior links for MPR and south direction for 2D Torus.

Figure 5.12 shows a graphical representation of these metrics for both the multipath ring and two-dimensional torus. It must be noted that η is only used

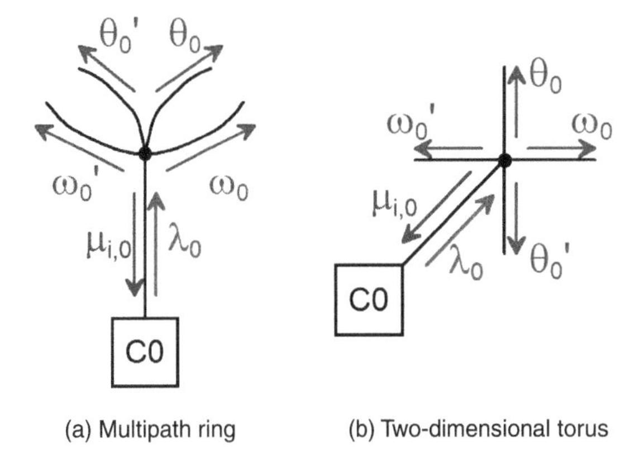

Figure 5.12 Cluster C0 metrics associated with the two analyzed topologies.

after the analysis is finished to correct the results by reducing the effects of overestimating the required throughput of the links. Also, because ω and θ are defined as the number of messages, they must be multiplied by the average output rate λ to get the actual throughput values.

Assuming a multipath ring network of size *N* and skip σ for every cluster *n*, the following equation stands:

$$\sum\nolimits_{0\leq i<N} \mu_{i,n} = \omega_{(n-1)modN} + \omega'_{(n+1)modN} + \theta_{(n-\sigma-1)modN} + \theta'_{(n+\sigma+1)modN} \tag{5.1}$$

where we consider $\mu_{n,n} = 0$. For example, in the case of the network shown in Figure 5.13 that has $N = 5$ and $\sigma = 1$, applying (5.1) results in the underdetermined system of Equations (5.2). The aim of the analysis is to solve these types of systems of equations in terms of ω, ω', θ, and θ'.

$$\begin{cases} \mu_{1,0} + \mu_{2,0} + \mu_{3,0} + \mu_{4,0} = \omega_4 + \omega'_1 + \theta_3 + \theta'_2 \\ \mu_{0,1} + \mu_{2,1} + \mu_{3,1} + \mu_{4,1} = \omega_0 + \omega'_2 + \theta_4 + \theta'_3 \\ \mu_{0,2} + \mu_{1,2} + \mu_{3,2} + \mu_{4,2} = \omega_1 + \omega'_3 + \theta_0 + \theta'_4 \\ \mu_{0,3} + \mu_{1,3} + \mu_{2,3} + \mu_{4,3} = \omega_2 + \omega'_4 + \theta_1 + \theta'_0 \\ \mu_{0,4} + \mu_{1,4} + \mu_{2,4} + \mu_{3,4} = \omega_3 + \omega'_0 + \theta_2 + \theta'_1 \end{cases} \tag{5.2}$$

A similar set of equations can be derived for the two-dimensional torus with N_R rows and N_C columns for every cluster *n*:

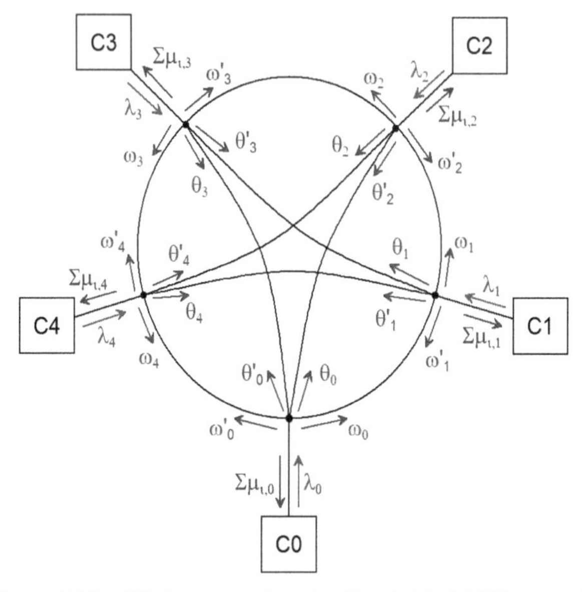

Figure 5.13 All the metrics in a size 5 and skip 2 MPR network.

$$\begin{aligned}\sum_{0\le i<N} \mu_{i,n} = {} & \omega_{\left\lfloor \frac{n}{N_C} \right\rfloor N_C + (n \bmod N_C - 1) \bmod N_C} \\ & + \omega'_{\left\lfloor \frac{n}{N_C} \right\rfloor N_C + (n \bmod N_C + 1) \bmod N_C} \\ & + \theta_{\left(\left\lfloor \frac{n}{N_C} \right\rfloor + 1\right) \bmod N_R \cdot N_C + n \bmod N_C} \\ & + \theta'_{\left(\left\lfloor \frac{n}{N_C} \right\rfloor - 1\right) \bmod N_R \cdot N_C + n \bmod N_C}\end{aligned} \tag{5.3}$$

Applying (5.3) to every node will also result in an underdetermined system of equations. A system of Equations based on (5.1) and (5.3) is underdetermined, having 4*N* unknown variables and only *N* equations for a network of size *N*, which makes solving it not a straightforward task. Various techniques can be used, such as QR decomposition [36], singular value decomposition [37], the pseudo-inverse, or least squares method [38]. However, in this particular case, a more efficient solution is to use superposition. Since the routing algorithm is static, the path followed by any message originating from any cluster is already known, and is independent of other clusters. As a result, the effects of traffic injected by one cluster can be accounted for while ignoring all the rest, which describes the superposition theorem. The first step for solving (5.2) is given in (5.4), where only the effects of cluster C0 are considered. After repeating this step for each cluster, the results can be combined and solution (5.5)-extracted.

$$\begin{cases} \omega_0 = \mu_{0,1} & \omega_0^{'} = \mu_{0,4} & \theta_0 = \mu_{0,2} & \theta_0^{'} = \mu_{0,3} \\ \omega_1 = 0 & \omega_0^{'} = 0 & \theta_0 = 0 & \theta_0^{'} = 0 \\ \omega_2 = 0 & \omega_0^{'} = 0 & \theta_0 = 0 & \theta_0^{'} = 0 \\ \omega_3 = 0 & \omega_0^{'} = 0 & \theta_0 = 0 & \theta_0^{'} = 0 \\ \omega_4 = 0 & \omega_0^{'} = 0 & \theta_0 = 0 & \theta_0^{'} = 0 \end{cases} \tag{5.4}$$

$$\begin{cases} \omega_0 = \mu_{0,1} & \omega_0^{'} = \mu_{0,4} & \theta_0 = \mu_{0,2} & \theta_0^{'} = \mu_{0,3} \\ \omega_1 = \mu_{1,2} & \omega_0^{'} = \mu_{1,0} & \theta_0 = \mu_{1,3} & \theta_0^{'} = \mu_{1,4} \\ \omega_2 = \mu_{2,3} & \omega_0^{'} = \mu_{2,1} & \theta_0 = \mu_{2,4} & \theta_0^{'} = \mu_{2,0} \\ \omega_3 = \mu_{3,4} & \omega_0^{'} = \mu_{3,2} & \theta_0 = \mu_{3,0} & \theta_0^{'} = \mu_{3,1} \\ \omega_4 = \mu_{4,0} & \omega_0^{'} = \mu_{4,3} & \theta_0 = \mu_{4,1} & \theta_0^{'} = \mu_{4,2} \end{cases} \tag{5.5}$$

In the case of such a small network, the resulting throughput values consist of only one μ parameter, and as such they are exact and no scaling with η is required. However, for bigger networks like the one shown in Figure 5.11(a), the solution is more complex, as in (5.6). Note that all μ parameters originating from the same cluster have been scaled with $1-\eta$. The routing scheme intervenes during each superposition step when computing the parameters. It is now clear that a static algorithm must be used otherwise it is impossible to apply superposition (adaptive routing requires the effect of all clusters to be considered, which contradicts the superposition assumptions).

$$\begin{cases} \omega_0 = (1-\eta)\left(\mu_{0,1} + \mu_{0,2} + \mu_{10,1} + \mu_{10,2}\right) + \mu_{12,1} + \mu_{0,4} \\ \omega_0^{'} = (1-\eta)\left(\mu_{0,12} + \mu_{0,11} + \mu_{3,12} + \mu_{3,11}\right) + \mu_{1,12} + \mu_{0,4} \\ \theta_0 = (1-\eta)\left(\mu_{0,3} + \mu_{0,4} + \mu_{0,5} + \mu_{0,6}\right) + \mu_{10,3} \\ \theta_0^{'} = (1-\eta)\left(\mu_{0,10} + \mu_{0,9} + \mu_{0,8} + \mu_{0,7}\right) + \mu_{3,10} \end{cases} \tag{5.6}$$

5.4 Energy-Delay Product

The energy-delay product [39] was used as an objective criterion for comparing the multipath ring with other commonly used topologies such as the mesh, binary tree, and their improved variants, inverse clustering and inverse clustering with mesh that is designed for low-power applications. Because, as done so far, the analysis is at a very abstract level, with no physical system built yet, the EDP estimation was also based on a set of simplified models that will be described in the following sections

5.4.1 Mathematical Derivation

The energy-delay product [39] (EDP) estimate for a data packet sent from cluster *i* to cluster *j* is defined as:

$$EDP_{i-j} = E_{i-j} * D_{i-j} \tag{5.7}$$

where E_{i-j} is the average energy required to send the packet and D_{i-j} is the delay of said packet. These two variables are calculated as shown in (5.8), where E_{router} and E_{link} are the average energies consumed in the router and link segment as the packet passes through and N_{router} and N_{link} are the number of routers and links in the entire path. The delay is influenced by three elements: D_{router}, the average delay through a router from one network port to another; D_{clust}, which is the delay from a cluster to its dedicated router, and D_{link}, the delay through a link segment. Figure 5.14 shows a schematic representation of all the energy and delay components in a path from cluster *i* to *j*.

$$\begin{aligned} E_{i-j} &= N_{router} * E_{router} + \Sigma E_{link} \\ D_{i-j} &= (N_{link} - 2) * D_{link} + 2 * D_{clust} + N_{router} * D_{router} \end{aligned} \tag{5.8}$$

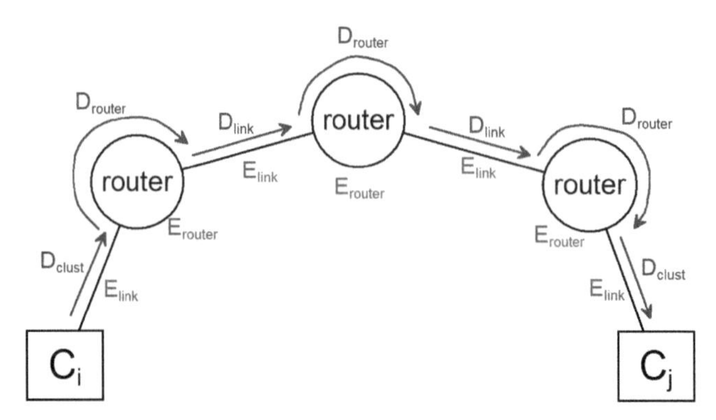

Figure 5.14 The energy and delay components that make up the path from cluster *i* to *j*.

Equation (5.7) is based on the original definition of the energy-delay product given in [39], but other works such as [41] have assigned more weight to the delay parameter by squaring or even cubing it. This method yields more accurate power estimations for performance-oriented systems, where a reduction in supply voltage produces a reduction with approximately the cube root of the power reduction in operating frequency (and, as a consequence, increased latency) [42]. For thoroughness, (5.9) will also be used in the energy-delay product estimation.

$$EDP_{i-j} = E_{i-j} * D_{i-j}^{3} \tag{5.9}$$

In all derivations, the index difference between cluster *i* and *j* defined as (5.10) will be used.

$$\Delta = |i - j| \tag{5.10}$$

Multipath Ring Network: For the MPR, the interior links are considered to be longer than the exterior ones and, as a consequence, the values for E_{link} are different and given by (5.11), where σ is the skip value.

$$E_{link} = \begin{cases} 1, & if\, external\, link \\ \sigma + 1, & if\, external\, link \end{cases} \tag{5.11}$$

N_{router}, N_{link}, and ΣE_{link} for every path between *i* and *j* are computed as

$$\begin{aligned} N_{router} &= \begin{cases} \left\lfloor \frac{\Delta}{\sigma+1} \right\rfloor + \Delta mod\,(\sigma+1) + 1, & \Delta \leq \frac{N}{2} \\ \left\lfloor \frac{N-\Delta}{\sigma+1} \right\rfloor + (N - \Delta)\, mod\,(\sigma+1) + 1, & \Delta > \frac{N}{2} \end{cases} \\ N_{link} &= N_{router} + 1 \\ \sum E_{link} &= \begin{cases} \left\lfloor \frac{\Delta}{\sigma+1} \right\rfloor (\sigma+1) + \Delta mod\,(\sigma+1) + 1, & \Delta \leq \frac{N}{2} \\ \left\lfloor \frac{\Delta}{\sigma+1} \right\rfloor (\sigma+1) + (N - \Delta)\, mod\,(\sigma+1) + 1, & \Delta > \frac{N}{2} \end{cases} \end{aligned} \tag{5.12}$$

Mesh Network: In this case, a square network is assumed, with the horizontal links physically shorter than the vertical ones by a factor of 4. For this reason, E_{link} is defined as:

$$E_{link} = \begin{cases} 1, & if\ horizontal\ link \\ 4, & if\ vertical\ link \end{cases} \tag{5.13}$$

which gives:

$$\begin{aligned} N_{router} &= \left\lfloor \frac{\triangle}{\sqrt{N}} \right\rfloor + \triangle mod \sqrt{N} + 1 \\ N_{link} &= N_{router} + 1 \\ \sum E_{link} &= \left\lfloor \frac{\triangle}{\sqrt{N}} \right\rfloor * 4 + \triangle mod \sqrt{N} + 1 \end{aligned} \quad (5.14)$$

Tree and inverse clustering: A small binary tree and inverse clustering network are shown in Figure 5.15. The inverse clustering network is described in [43] and is essentially a rearrangement of the binary tree so that clusters that are farther away from each other are connected with shorter paths. This has the effect of reducing the delay and energy through these paths since long distance traveling packets no longer need to pass through a large number of routers, and in the case of CMOS, IC implementations can be routed through lower metal layers.

Figure 5.15 also shows the proportionality factors for E_{link}. For the binary tree, the lower the level, the shorter the links are assumed to be, a trait which is reversed for the inverse clustering. Under these assumptions, (5.15) are the equations used for calculating all necessary parameters for the binary tree.

$$\begin{aligned} N_{router} &= 2 * \lfloor \log_2 \triangle \rfloor + 1 \\ N_{link} &= N_{router} + 1 \\ \sum E_{link} &= 2^{\lfloor \log_2 \triangle \rfloor + 2} - 2 \end{aligned} \quad (5.15)$$

Inverse clustering with mesh: Inverse clustering with mesh is a hybrid architecture proposed in [43] that, as the name suggests, uses a mesh to connect neighboring clusters and the inverse clustering for longer paths. Figure 5.16 shows an example of such a topology for a network of 16 clusters. The corresponding E_{link} proportionality factors are also indicated on the links.

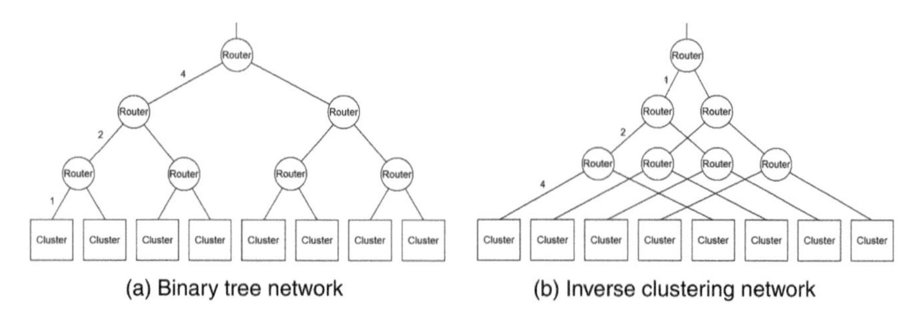

Figure 5.15 Comparison between binary tree and inverse clustering networks of size 8. The numbers on the links indicate proportionality factors between each corresponding E_{link}.

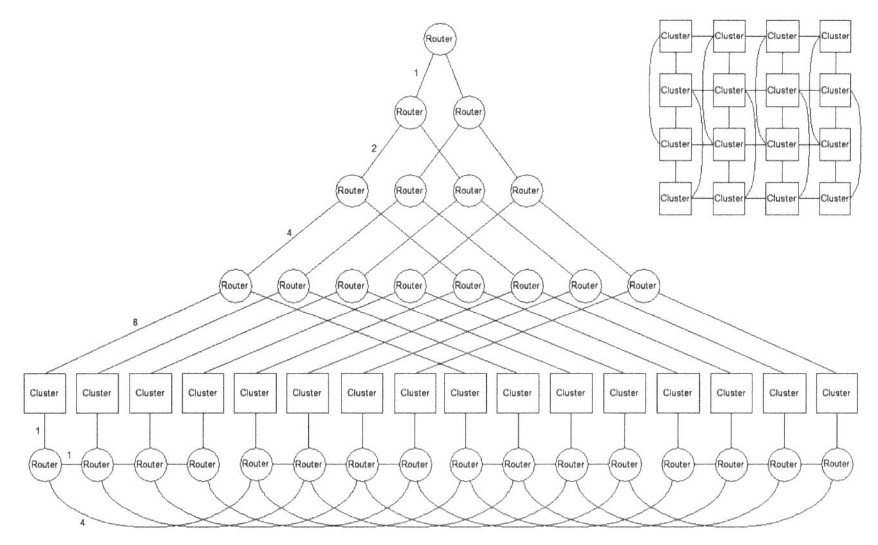

Figure 5.16 The inverse clustering with mesh network topology and the associated E_{link} proportionality factors. The inset shows the same network arranged as a mesh with the connections in the first inverse cluster layer visible.

Finding an exact equation for computing the three parameters of interest is difficult because of the hybrid nature of the architecture and so, for the simulations, all parameters were manually calculated by analyzing every path between cluster i and j and choosing the shortest one.

5.4.2 Energy-Delay Product Estimation

Figure 5.17(a) shows the energy-delay product estimations for a network of size 32 arranged as a function of Manhattan distance according to (5.7). For small distances, the multipath ring has the smallest estimated EDP for all skip values. Tree and inverse clustering with mesh have an almost identical EDP, with the latter behaving marginally worse. As the distance increases, MPR with skip 1 becomes the best choice out of the three MPR topologies, with mesh becoming slightly more advantageous for large distances. Nevertheless, MPR with skips 2 and 3, which have almost the same EDP, have a smaller maximum distance (network diameter) than all the other topologies and remain the attractive solution.

Figure 5.17(b) illustrates the EDP estimates using (5.9). The same characteristics are visible, but now the mesh topology has a greater advantage over the multipath ring with skip 1 for large distances. Nevertheless, the MPR

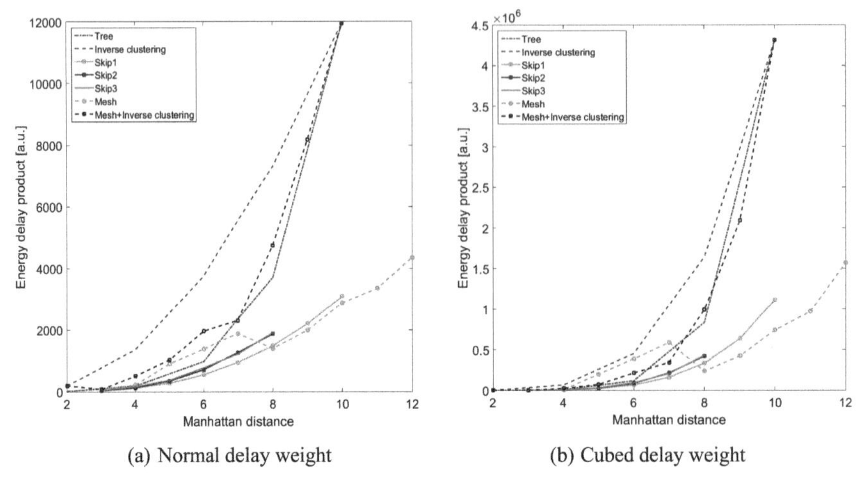

(a) Normal delay weight

(b) Cubed delay weight

Figure 5.17 EDP estimations for different network topologies of 32 clusters vs. Manhattan distance for different delay weights.

topologies remain attractive for small to medium distances due to their very low EDP and small diameters. Since data locality is assumed, the advantages that the mesh topology exhibits for large distances have very low probabilities to manifest itself.

Figure 5.18 shows the estimated EDP according to (5.7) as a function of Δ, the index difference, for a multipath ring topology of size 32. Data was collected from a Monte Carlo simulation with 1000 iterations. The symmetric nature of the architecture can be seen in the shape of the EDP. The peak represents the worst case when $\Delta = (N - 1)/2$, i.e., when the source and destination cluster are on opposite sides of the network. According to the definition, packets traveling through the exterior links of the MPR are counted by ω and ω' and all of them have the EPR value corresponding to $\Delta = 1$. The interior traffic expressed by θ and θ' depends on the skip value σ and has an EDP corresponding to $\Delta = \sigma + 1$. Following these rules, the global EDP can be computed for every one of the 1000 iterations with:

$$EDP = \sum_{i=0}^{N-1} \left(\omega_i + \omega_i^{'}\right) * EDP_1 + \sum_{i=0}^{N-1} \left(\theta_i + \theta_i^{'}\right) * EDP_{\sigma+1} \tag{5.16}$$

where N is the number of clusters.

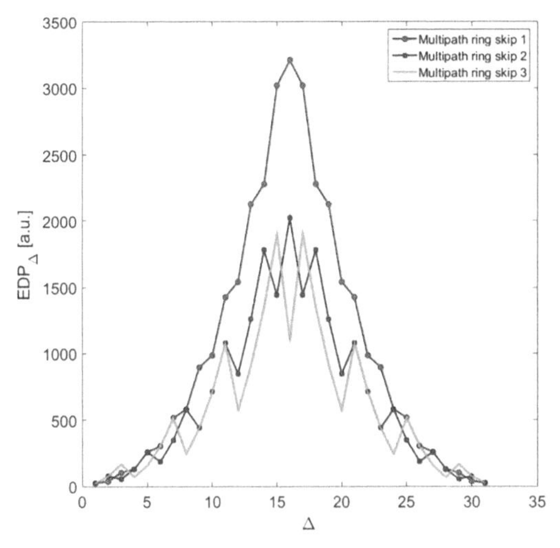

Figure 5.18 Energy-delay product as a function of index difference in the multipath ring of size 32.

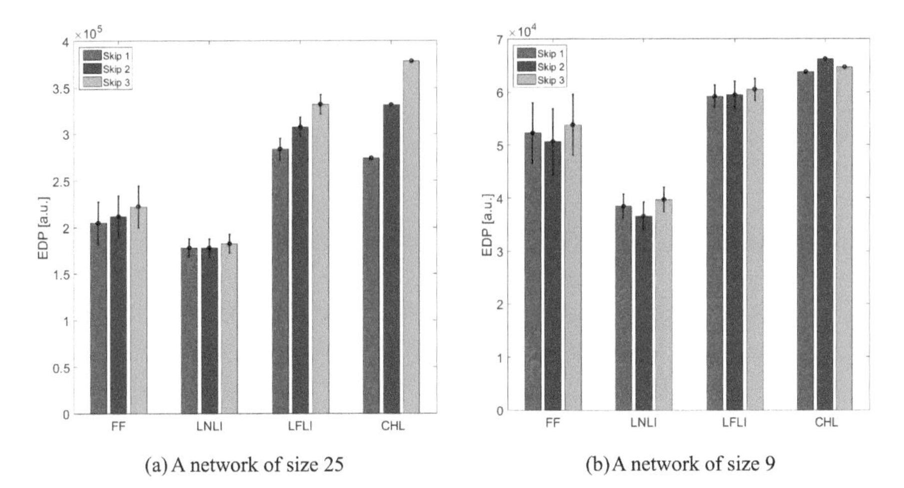

(a) A network of size 25 (b) A network of size 9

Figure 5.19 Global energy-delay products for all communication schemes with traffic resulting from random neuron placements.

The average of the 1000 global EDP estimates are shown in Figure 5.19, with the error bars representing one standard deviation. Results indicate that the communication scheme, i.e., traffic distribution, has a strong effect on the energy-delay product regardless of network size. It is even more prominent for a large number of small clusters since the number of intracluster connections is small, and there is more strain on the network. In this case, EDP can vary with up to 50% for MPR with skip 3 between LNLI and CHL.

5.5 Conclusions

The multipath ring (MPR) topology is proposed as an efficient dataflow architecture for neuron-to-neuron communication in a clustered spiking neuron network simulator to be implemented on FPGA. When compared with the two-dimensional torus, an architecture commonly employed for multicore platforms improved throughput capacity was found, along with the possibility of traffic shaping, or redirecting through specific links. By configuring MPR parameters such as skipping distance, the topology can be adapted for specific network sizes and to maintain very efficient traffic balance. The MPR is capable of handling different neuron communication schemes as well, which makes it suitable for supporting the continuous changes in traffic patterns due to learning processes in the neural network.

As a method for characterizing the energy efficiency of the multipath ring network topology, the energy-delay product was estimated and compared with other network types, such as the low-power inverse-clustering with mesh. The observed values indicate an energy-efficient topology that has the added benefit of small network diameter and physical implementation simplicity.

Strain on the network was reduced using two physically separate communication layers, one for the neuron-to-neuron data, for which the MPR was proposed, and one for configuration and input/output data. For the former, a binary-tree topology was adapted, with a channel routing protocol for downstream traffic. In the case of upstream traffic, no routing decisions have to be made apart from arbitration. Overall, the routers can be kept simple and do not require routing tables.

References

[1] W. Gerstner and W.M. Kistler, *Spiking Neuron Models*. Cambridge University Press, 2002.

[2] W. Teka, T. Marinov, and F. Santamaria, "Neuronal spike timing adaptation described with a fractional leaky integrate-and-fire model," *PLoS Computational Biology*, vol. 10, p. e1003526, Mar 2014.

[3] E.M. Izhikevich, "Simple model of spiking neurons," *IEEE Transactions on Neural Networks*, vol. 14, pp. 1569–1572, Nov 2003.

[4] A.L. Hodgkin and A.F. Huxley, "A quantitative description of membrane current and its application to conduction and excitation in nerve," *Journal of Physiology*, vol. 117, no. 4, pp. 500–544, 1952.

[5] R. Wang, T. J. Hamilton, J. Tapson, and A. van Schaik, "An fpga design framework for large-scale spiking neural networks," in *2014 IEEE International Symposium on Circuits and Systems (ISCAS)*, pp. 457–460, June 2014.

[6] U. Farooq, Z. Marrakchi, and H. Mehrez, *Tree-based Heterogeneous FPGA Architectures*. Springer, 2012.

[7] R. Miikkulainen, "Neuroevolution," in *Encyclopedia of Machine Learning* (C. Sammut and G. I. Webb, eds.), pp. 716–720, Boston, MA: Springer US, 2010.

[8] G.J. Christiaanse, A. Zjajo, C. Galuzzi, and R. van Leuken, "A real-time hybrid neuron network for highly parallel cognitive systems," in *2016 38th Annual International Conference of the IEEE Engineering in Medicine and Biology Society (EMBC)*, pp. 792–795, Aug 2016.

[9] J. Hofmann, A. Zjajo, C. Galuzzi, and R. van Leuken, "Multi-chip dataflow architecture for massive scale biophysically accurate neuron simulation," in *2016 38th Annual International Conference of the IEEE Engineering in Medicine and Biology Society (EMBC)*, pp. 5829–5832, Aug 2016.

[10] G.J. Christiaanse, "Real-time hybrid neuron network for highly parallel cognitive systems." MSc Thesis Delft University of Technology, The Netherlands, 2016.

[11] J.D. Gruijl, P. Bazzigaluppi, M.T.G. de Jeu, and C.I. de Zeeuw, "Climbing fiber burst size and olivary sub-threshold oscillations in a network setting," *PLoS Computing Biology*, vol. 12, p. e1002814, Dec 2012.

[12] E.M. Izhikevich, "Which model to use for cortical spiking neurons?," *IEEE Transactions on Neural Networks*, vol. 15, pp. 1063–1070, Sep 2004.

[13] S.B. Furber, F. Galluppi, S. Temple, and L.A. Plana, "The spinnaker project," *Proceedings of the IEEE*, vol. 102, pp. 652–665, May 2014.

[14] S.B. Furber, D.R. Lester, L.A. Plana, J.D. Garside, E. Painkras, S. Temple, and A.D. Brown, "Overview of the spinnaker system architecture," *IEEE Transactions on Computers*, vol. 62, pp. 2454–2467, Dec 2013.

[15] B.V. Benjamin, P. Gao, E. McQuinn, S. Choudhary, A.R. Chandrasekaran, J.M. Bussat, R. Alvarez-Icaza, J.V. Arthur, P.A. Merolla,

and K. Boahen, "Neurogrid: A mixed-analog-digital multichip system for large-scale neural simulations," *Proceedings of the IEEE*, vol. 102, pp. 699–716, May 2014.

[16] P. Merolla, J. Arthur, R. Alvarez, J.M. Bussat, and K. Boahen, "A multicast tree router for multichip neuromorphic systems," *IEEE Transactions on Circuits and Systems I: Regular Papers*, vol. 61, pp. 820–833, Mar 2014.

[17] P.A. Merolla, J.V. Arthur, R. Alvarez-Icaza, A.S. Cassidy, J. Sawada, F. Akopyan, B.L. Jackson, N. Imam, C. Guo, Y. Nakamura, B. Brezzo, I. Vo, S.K. Esser, R. Appuswamy, B. Taba, A. Amir, M.D. Flickner, W. P. Risk, R. Manohar, and D.S. Modha, "A million spiking-neuron integrated circuit with a scalable communication network and interface," *Science*, vol. 345, no. 6197, pp. 668–673, 2014.

[18] M. Beuler, A. Tchaptchet, W. Bonath, S. Postnova, and H.A. Braun, "Real-time simulations of synchronization in a conductance-based neuronal network with a digital fpga hardware-core," in *Artificial Neural Networks and Machine Learning – ICANN 2012: 22nd International Conference on Artificial Neural Networks, Lausanne, Switzerland, September 11–14, 2012, Proceedings, Part I* (A. E. P. Villa, W. Duch, P. Erdi, F. Masulli, and G. Palm, eds.), pp. 97–104, Berlin, Heidelberg: Springer Berlin Heidelberg, 2012.

[19] H.A. Braun, M T. Huber, M. Dewald, K. Schfer, and K. Voigt, "Computer simulations of neuronal signal transduction: The role of nonlinear dynamics and noise," *International Journal of Bifurcation and Chaos*, vol. 08, no. 05, pp. 881–889, 1998.

[20] M. Beuler, A. Krum, W. Bonath, and H. Hillmer, "Nepteron processor for realtime computation of conductance-based neuronal networks," in *2017 Euromicro Conference on Digital System Design (DSD)*, pp. 78–85, Aug 2017.

[21] H. Shayani, P.J. Bentley, and A.M. Tyrrell, "Hardware implementation of a bio-plausible neuron model for evolution and growth of spiking neural networks on fpga," in *2008 NASA/ESA Conference on Adaptive Hardware and Systems*, pp. 236–243, Jun 2008.

[22] K.L. Rice, M.A. Bhuiyan, T.M. Taha, C.N. Vutsinas, and M.C. Smith, "Fpga implementation of izhikevich spiking neural networks for character recognition," in *2009 International Conference on Reconfigurable Computing and FPGAs*, pp. 451–456, Dec 2009.

[23] M. Ambroise, T. Levi, Y. Bornat, and S. Saighi, "Biorealistic spiking neural network on fpga," in *2013 47th Annual Conference on Information Sciences and Systems (CISS)*, pp. 1–6, Mar 2013.

[24] E. Cota, A. de Morais Amory, and M.S. Lubaszewski, *Reliability, Availability and Serviceability of Networks-on-Chip*. Springer, 2012.

[25] J. Duato, S. Yalamanchili, and N. Lionel, *Interconnection Networks: An Engineering Approach*. San Francisco, CA, USA: Morgan Kaufmann Publishers Inc., 2002.

[26] J.-S. Shen and P.-A. Hsiung, *Dynamic Reconfigurable Network-on-Chip Design: Innovations for Computational Processing and Communication*. Hershey, PA: Information Science Reference - Imprint of: IGI Publishing, 2010.

[27] T. Pionteck, C. Albrecht, and R. Koch, "A dynamically reconfigurable packetswitched network-on-chip," in *Proceedings of the Design Automation Test in Europe Conference*, vol. 1, pp. 8, Mar 2006.

[28] M. B. Stensgaard and J. Spars, "Renoc: A network-on-chip architecture with reconfigurable topology," in *Second ACM/IEEE International Symposium on Networks-on-Chip (nocs 2008)*, pp. 55–64, Apr 2008.

[29] E. Fiesler, "Neural network topologies," 1996.

[30] R. Miikkulainen, "Topology of a neural network," in *Encyclopedia of Machine Learning* (C. Sammut and G. I. Webb, eds.), pp. 988–989, Boston, MA: Springer US, 2010.

[31] J.J. Hopfield, "Neural networks and physical systems with emergent collective computational abilities," *Proceedings of the National Academy of Sciences*, vol. 79, no. 8, pp. 2554–2558, 1982.

[32] H.K. Hartline, H.G. Wagner, and F. Ratliff, "Inhibition in the eye of limulus," *The journal of general physiology*, vol. 5, no. 39, pp. 651–673, 1956.

[33] R. Coultrip, R. Granger, and G. Lynch, "A cortical model of winner-take-all competition via lateral inhibition," *Neural Networks*, vol. 5, no. 1, pp. 47–54, 1992.

[34] D.J. Watts and S.H. Strogatz, "Collective dynamics of 'small-world' networks," *Nature*, vol. 393, no. 6684, pp. 440–442, 1998.

[35] Xilinx. http://www.xilinx.com.

[36] "21. underdetermined systems," in *Accuracy and Stability of Numerical Algorithms*, pp. 407–414.

[37] E.J. Ientilucci. http://www.cis.rit.edu/~ejipci/Reports/svd.pdf.

[38] C. Lawson and R. Hanson, *Solving Least Squares Problems*. Society for Industrial and Applied Mathematics, 1995.

[39] M. Horowitz, T. Indermaur, and R. Gonzalez, "Low-power digital design," in *Proceedings of 1994 IEEE Symposium on Low Power Electronics*, pp. 8–11, Oct 1994.

[40] X. XPE. https://www.xilinx.com/products/technology/power/xpe.html.

[41] D.M. Brooks, P. Bose, S.E. Schuster, H. Jacobson, P.N. Kudva, A. Buyuktosunoglu, J. Wellman, V. Zyuban, M. Gupta, and P.W. Cook, "Power-aware microarchitecture: design and modeling challenges for next-generation microprocessors," *IEEE Micro*, vol. 20, pp. 26–44, Nov 2000.

[42] M.J. Flynn, P. Hung, and K.W. Rudd, "Deep submicron microprocessor design issues," *IEEE Micro*, vol. 19, pp. 11–22, Jul 1999.

[43] V. George, H. Zhang, and J. Rabaey, "The design of a low energy fpga," in *Proceedings. 1999 International Symposium on Low Power Electronics and Design (Cat. No.99TH8477)*, pp. 188–193, Aug 1999.

[44] Accellera. http://www.eda.org/activities/working-groups/systemc-synthesis.

6

A Hierarchical Dataflow Architecture for Large-Scale Multi-FPGA Biophysically Accurate Neuron Simulation

He Zhang

Delft University of Technology, Delft, The Netherlands

The scalable simulation of neuron communication requires a large amount of computing resources. The high throughput of data poses significant challenges on the interconnect network. In this chapter, we propose an efficient, hierarchical dataflow architecture for large-scale biophysically accurate multichip implementation of the neural network. The network is characterized in terms of the topology, routing, and flow control. To find the efficient network structure, we perform analysis of the throughput for the different network with different traffic patterns based on the hopcount and bandwidth. The results indicate that the multicast in mesh topology offers 33% improvement in comparison with unicast. Based on the interconnect router architecture, we built a cycle accurate simulator in SystemC.

6.1 Introduction

Providing highly flexible connectivity is a major architectural challenge for hardware implementation of reconfigurable neural networks. The neural networks are formed in the different shape of connections, which may influence the performance for the hardware router connection topology. To simulate the behaviour of a highly parallel cognitive system, the communication traffic within the interconnection rises exponentially by the scalability of neurons. The previous work [1] implements maximum 1188 Hodgkin–Huxley model neurons in one field-programmable gate array (FPGA), which are locally connected to a binary tree network with routers. The FPGA communicates with the PC through an Ethernet port. To increase the number of implemented neurons beyond single FPGA, a multi-FPGA system is considered as a solution. The system has the ability to handle the three types of real-time packets, i.e., the initialization, the internal communication packets, and outputs. A multi-FPGA system described in this chapter is capable of simulating a large number of neurons in real time. The design can be configured with the scale size, flit size, and virtual channel unicast/multicast strategy for different neuron connection topologies. Three traffic patterns, uniform traffic, fully connected traffic, and all-to-one traffic are generated to test the performance of the network. We also consider the limitation of Xilinx FPGA select IO port to construct the serializer/deserializer interface.

This chapter is organized as follows: Section 6.2 summarizes the interconnection network and SerDes interface and discusses the performance with different constraints. Section 6.3 gives the implementation of the router in SystemC simulation and SerDes interface in VHDL. Section 6.4 discusses the simulation results. Section 6.5 concludes the chapter.

6.2 The System Overview

The system can be considered as a hierarchical structure consisting of a mesh network of the FPGAs (Figure 6.1), and within each FPGA, a tree network connecting clusters and neuron cores in the clusters (Figure 6.2). Field-programmable gate arrays (FPGAs), although slower than ASIC, provide sufficient parallelism and performance to allow for real-time and even hyper-real-time neuron network simulation [1–3]. Various neuron models have been implemented on FPGAs, e.g., integrate-and-fire model [4], the Izhikevich model [5], or extended Hodgkin–Huxley model [2].

6.2.1 Mesh Topology

Mesh topology with multicast communication offers improved performance in comparison to fat-tree, point-to-point, and shared bus communication for neural networks in completed and random connection [6]. The mesh router has dual ports connecting to the neighbor router for full duplex communication (Figure 6.1). Inside each FPGA is a tree-based architecture, where the clusters of neurons are located at the leaves. The data generated from the neurons are up-spread to the root of the tree and spread to other FPGAs using the mesh router.

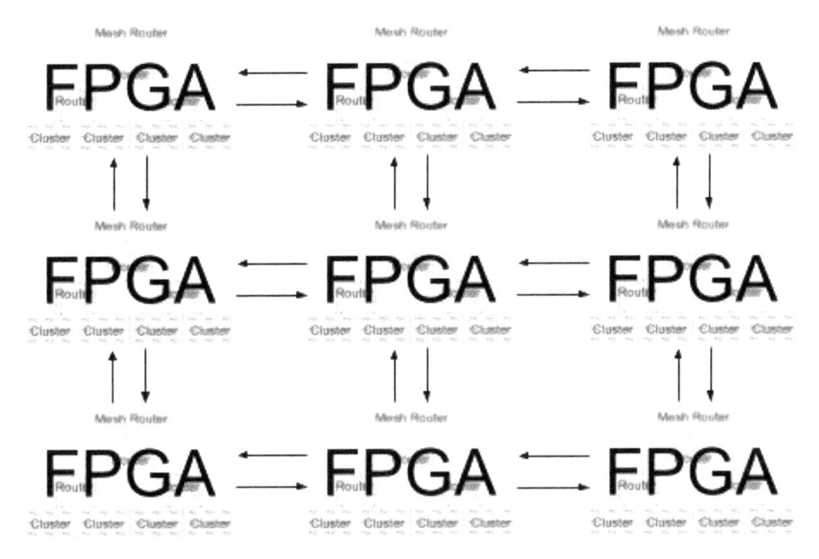

Figure 6.1 The mesh connection.

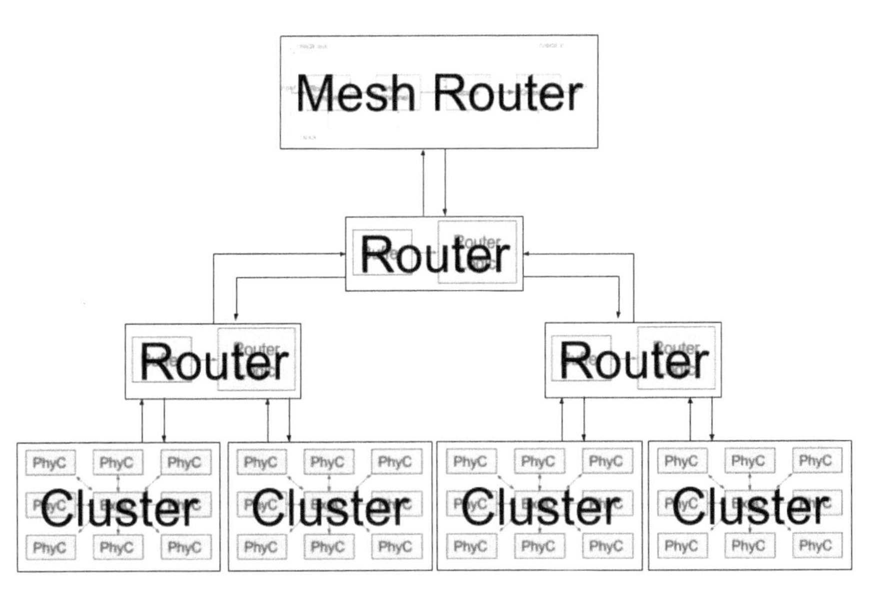

Figure 6.2 Tree structure within one FPGA.

6.2.2 The Routers

In the simulator, both unicast and multicast routings supported with different allocator structure in the mesh routers. This structure has five main parts, namely the input buffer, the switch allocator, the virtual channel allocator, the crossbar switch, and the route computation, as shown in Figure 6.3. The data path is set in a four-stage pipeline through these modules.

The point-to-point unicast routing algorithm uses the dimension order routing algorithm. The multicast routing algorithm we used is tree-based on the dimension order routing (DOR) [7]. The routing path is the collection of all paths in the dimension order for unicast with only one route path for each node. The multicast DOR routing is also deadlock-free. The routing destinations are pre-computed and stored in the routing table, which is checked at the input buffer. The flow is controlled by the credit-based backpressure, i.e., the number of valid buffer slots of each virtual channel in the downstream is counted and updated by the credit signal sent in when new buffer slots are made available. The allocator offers virtual channel selection and switch allocation. Channel allocator is based on round-robin sequence. Each head flit is asked for one output channel, and the body or tail flit keep the same virtual channel as the head. The switch allocator allocates the output port for the flits, which has taken the virtual channel. The allocator also uses Round-robin strategy to arbitrate when contentions occur in one output. The crossbar switch is controlled by the allocation decision.

The router structure: The router structure has five main parts: input buffer, switch allocator, virtual channel allocator, crossbar switch, and processing unit interface.

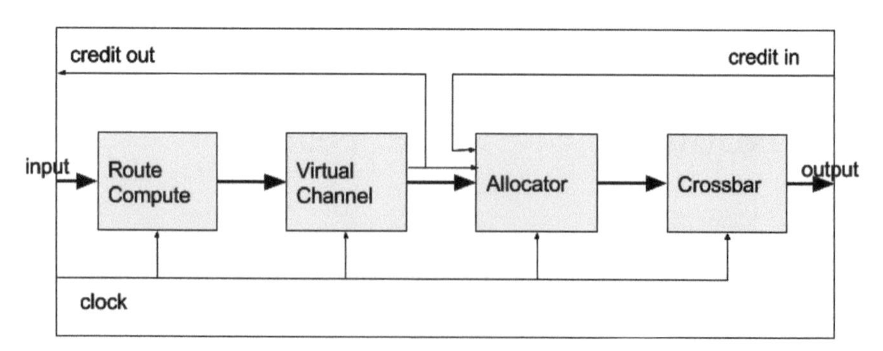

Figure 6.3 Router structure schematic.

Virtual channel: Virtual channels are used to buffer the packets that are in contention for the same physical channel and, consequently, increase the throughput. A large number of virtual channels require larger FIFOs and complex arbiter and allocator to route the packets.

Buffer: The routing table is checked at the buffer write stage. A mesh topology connection is implemented with the XY routing algorithm. The flowchart for the arbiter is shown in Figure 6.4. Each input port is connected to an FIFO, whose size has less influence on the packet delay comparing to network load [8].

The back-propagating signal can be transferred within separate physical channel or combined with flits. However, a trade-off exists between transfer latency and port resources. For separate channels, the credit can be transferred within one cycle latency in parallel (it will cost four cycles to transfer multiplexing channel with data by serial). The additional latency in the back-propagating transmission decreases the throughput of the network, or need more virtual channels to multiplex using the physical channels.

Channel allocator is based on the Round-robin sequence. Each head flit is asked for one output channel and the body or tail flit keep the same virtual channel as the head.

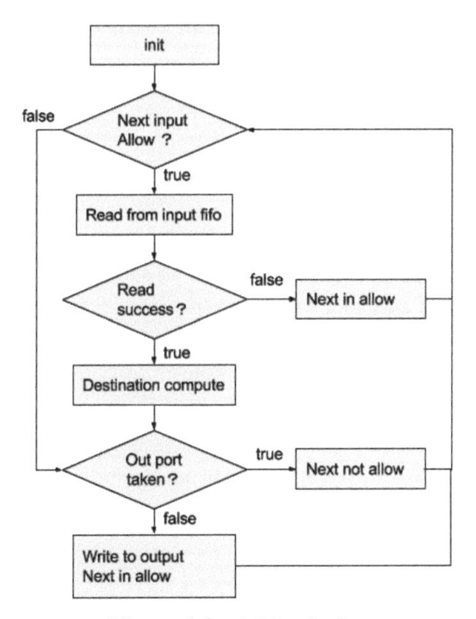

Figure 6.4 Arbiter logic.

Switch: The switch allocator allocates the output port for the flits, which has taken the virtual channel. The allocator also uses the Round-robin strategy to arbitration when the contention occurs in one output.

Crossbar switch: The crossbar switch is the structure for the dataflow from input to output. It is controlled by the allocation decision.

6.2.3 The Clusters

The communication architecture is responsible for initializing and delivering the requested packet containing the dendrite or axon potential to the calculation architecture and communicating to an external interface outside of the proposed design. It contains three levels: interface bridge, routing network, and cluster controllers [1]. The cluster controller is responsible for relating new values to the calculation architecture when requested and storing and routing their responses while also storing the responses of other physical cells (PhyCs). Information between clusters and the interface bridge is passed through a balanced tree network. The two data streams, axon hillock response and dendrite response, are transferred to the top and to all, respectively. The interface bridge is capable of communicating to and from the (virtual) boundary pins of the design and the routing network. At the interface, two kinds of packets are transferred, the initial packet and the output packet.

6.2.4 Hodgkin–Huxley Cells

For electrochemically accurate neuron modeling, which is the focus of our study, the conductance-based multi-compartment Hodgkin–Huxley model [9] is required. Biophysically accurate models of biological systems, such as the ones using the Hodgkin–Huxley formalism, are composed mostly of a set of computationally challenging differential equations often implementing an oscillatory behavior. If the interconnectivity between oscillating neurons is also modeled (e.g., gap junctions, input integrators, and synapses), the cells become coupled oscillators. Consequently, all neuron states need to be completely updated at each simulation step to retain correct functionality. As a result, cycle-accurate, transient simulator is necessary. The above difficulties in associated HH models and multi-compartmental models with complex connections, in conjunction with biophysically plausible neuron network sizes, pose significant challenges especially when using conventional computing machines.

The calculation architecture is based on the extended version of the HH model, which describes the inferior olive neurons (ION) as a multiple-compartmental cell. Each compartment within the model holds a state potential, which is updated every simulation step. The states depend on the acquired chemical currents, and the coupling effect defined by the dendritic gap-junction couplings present between the neurons. To increase efficiency, the calculations are grouped into two parallel processes: the first group, which performs calculations independent of the neighboring neurons housed by soma and axon (hillock) compartments, and the second group, which depends on the coupling effect on the dendrite compartment [1].

6.3 The Communication Architecture

The communication architecture is responsible for initializing and delivering requested packet containing the dendrite or axon potential to the calculation architecture, and communication to an external interface outside of the proposed design. It contains three levels: interface bridge, routing network, and cluster controllers [1]. The cluster controller is responsible for relating new values to the calculation architecture when requested and storing and routing their responses while also storing the responses of other PhyCs. Information between clusters and the interface bridge is passed through a balanced tree network. The two data streams, axon hillock response and dendrite response, are transferred to the top and to all, respectively. The interface bridge is capable of communicating to and from the (virtual) boundary pins of the design and the routing network. At the interface, there will be two kinds of packets transferred, initial packet and output packet.

Considering the network topology and cell communication strategy, in [3], a re-configurable multichip architecture is presented, which reduces the communication cost to linear growth. For multi-FPGAs synchronization, a control module is set in one special FPGA connected to slave FPGAs, which are connected in ring topology. In [1], the physical neurons (PhyCs) are time-division multiplexed to operate for several simulation neurons (SimCs). Each SimC within the design needs to calculate and communicate simulated responses to their neighbors and the axon. The calculation and communication are considered separately.

The bandwidth requirement is specified by the data throughput. In [1], the message, as listed in Tables 6.1 and 6.2, consists of a 32/64 bits data, an address covering all the neurons and 2 bits of data type, which indicate the usage of data value. So, the message size is $64 + \log_2 N + 2$. The average

Table 6.1 Input init data vector [1]

Used by	Name	Bits
Bridge	*cluster_init_clus*	*calc_bits(#Clusters_per_FPGA −1)*
PCC	*cluster_init_type*	3
	cluster_init_adr	*calc_bits(#SimC −1)*
	cluster_init_adr2	*calc_bits(max(#params −1, #Connections))*
	cluster_init_data	32/64

$calc\_bits(x) = 1 + log_2(max(x, 1))$

Table 6.2 Output data vector [1]

Used by	Name	Bits
Output	*cluster_out_type*	2
	cluster_out_adr	*calc_bits(#SimC −1)*
	cluster_out_data	32/64

$calc\_bits(x) = 1 + log_2(max(x, 1))$

bandwidth of each FPGA is 40,000 × N × (64 + $\log_2$N + 2) bit/s. Assuming the number of neuron N = 1000, the total bandwidth will be more than 3 Gbit/s. The peak bandwidth is achieved when the communication is in cycle accurate, which means every cycle 64 + $\log_2$N + 2 bit of data and address need to be transferred. The peak bandwidth will be 7.6 Gbps if we assume 10 ns clock frequency. The message latency here is defined as the cycles it takes to transfer the message from the source node to the destination node. The message latency has the relation with the calculation cycles that the total operating cycles should be within 5000 (for 10 ns clock period).

The packet model: The message as shown in Table 6.3 consists of an address, and axon or dendrite potential data, which are in the same format and width. To make the network suitable for the pseudo-random connection matrix, the destination is determined by lookup in the routing table generated at

Table 6.3 Date and signal contained in packet model

Name	Width
data	32/64
type	2
address	logN
direction	5
flit type	2
virtual channel No.	2

the initial section. The data in the packet designates the voltage for axon or dendrite, which has 32 and 64 bit widths for float type and double type, respectively. The value of *type* indicates which type the data stand for; *adr* is source address of the packet for checking lookup table; h_t signal has three values for flit flow control, where 01 stands for head, 11 stands for body flit, and 10 for the tail flit; vc_{num} describes virtual channel distribution during the transfer, which can be allocated to different virtual channels at the input port.

Topology analysis: Providing highly flexible connectivity is a major architectural challenge for hardware implementation of reconfigurable neural networks. The neural network is formed in different shapes of connections, which may influence the performance for the hardware router connection topology. To examine the relationship between the allocation between neural network and hardware resources, the analysis is based on the several neuron distribution and routing topologies with three packet-based communication methods (unicast, multicast, and broadcast).

Throughput depends on the parallelism in the architecture and the interconnect bandwidth [10]. In order to compare different architectures that provide the same throughput and end-to-end delay, an effective bandwidth as the actual communication bandwidth or throughput is defined as [10]:

$$BW\mathit{eff} = \frac{\sum num\_link}{TotalHopCount} \times f \times U \quad (6.1)$$

where *num* link is the number of links between each node, *TotalHopCount* is the average total distance traversed by each packet (going to all destinations), measured in the number of hops, f is the operation frequency, and U is the utilization of the architecture [6]. We assume that the router works in the realistic condition, which transfers all input data. All circuits operate at the same clock frequency. The cost is compared by the width of channels.

$$LinkWidth = \sum num\_link \quad (6.2)$$

The cost-performance ratio *R* is:

$$R = \frac{BW\mathit{eff}}{LinkWidth} \quad (6.3)$$

A n-ary mesh comprises n processors and n routers arranged in a $\sqrt{n}$ by $\sqrt{n}$ mesh. The total number of links in 2D mesh is:

$$LinkWidth_{mesh} = 2\sqrt{n}\left(\sqrt{n} - 1\right) \quad (6.4)$$

The binary tree topology has n neurons at the leaves, and $n - 1$ routers compose the branches. The fat tree has the same topology but the links increase to the root that the number of links in each level remains the same of n. The total number of links in binary tree is:

$$LinkWidth_{tree} = 2\,(n-1) \tag{6.5}$$

and in the fat tree is:

$$LinkWidth_{fatTree} = nlog_2 n \tag{6.6}$$

Traffic Pattern Analysis: Hopfield Neural Network
Hopfield network is the network with all-to-all connection to other neurons, and designates the upper bound on the network traffic. Here, we emulate Hopfield network on 2D mesh and tree topology. The average distance between two nodes in 2D mesh and tree is $(2/3)\sqrt{n}$ and $[2n(\log_2 n - 1) + 2]/(n-1)$, respectively. The total hop count for one neuron packet is the sum of one neuron to the others:

$$\begin{aligned} TotalHopCount^{mesh}_{unicast} &= (n-1)\,\tfrac{2}{3}\sqrt{n} \\ TotalHopCount^{tree}_{unicast} &= 2n(log_2 n - 1) + 2 \end{aligned} \tag{6.7}$$

So, the throughput is:

$$\begin{aligned} BWeff^{mesh}_{unicast} &= \tfrac{3}{\sqrt{n}+1} = O\!\left(n^{-\frac{1}{2}}\right) \\ BWeff^{tree}_{unicast} &= \tfrac{n-1}{n(log_2 n-1)+1} = O\!\left(\tfrac{1}{log_2 n}\right) \\ BWeff^{fatTree}_{unicast} &= \tfrac{nlog_2 n}{2n(log_2 n-1)+2} = O(1) \end{aligned} \tag{6.8}$$

Because of the completely connection of Hopfield network, the multicast and broadcast are the same. The total hop count of broadcast in mesh topology is the number of edges in spanning tree, which is $n - 1$ and in tree topology, it is $2(n - 1)$. Consequently,

$$\begin{aligned} BWeff^{mesh}_{broadcast} &= \tfrac{2\sqrt{n}}{\sqrt{n}+1} = O(1) \\ BWeff^{tree}_{broadcast} &= O(1) \\ BWeff^{fatTree}_{broadcast} &= \tfrac{nlog_2 n}{2(n-1)} = O(log_2 n) \end{aligned} \tag{6.9}$$

The cost for each topology is as follows:

$$\begin{aligned} Cost_{mesh} &= 2\sqrt{n}\,(\sqrt{n}-1) = O(n) \\ Cost_{tree} &= 2\,(n-1) = O(n) \\ Cost_{fatTree} &= nlog_2 n = O(nlog_2 n) \end{aligned} \tag{6.10}$$

The summary of the complexity of different topology implementing Hopfield network is listed in Table 6.4.

The Randomly Connected (RNDC) Neural Network: In realistic scenarios, the neurons have random exponential local connectivity [11]. The probability of connection within two neurons follows negative correlation with their distance:

$$p(a,\,b) = \frac{C}{2\pi\lambda^2} e^{-D(a,\,b)/\lambda} \tag{6.11}$$

where C is the average connection per neuron, $D(a,b)$ is the Euclidean distance between neurons a and b, and λ is a spatial connectivity constant reflecting the average traversing distance. Here, we map the neurons to the mesh topology by the distance. The average distance is:

$$avgDist = \frac{C}{2\pi\lambda^2} \int De^{-D/\lambda} dD = 2\lambda C \tag{6.12}$$

where $D = \sqrt{x^2 + y^2}$

The unicast total average hop count is:

$$\begin{aligned} TotalHopCount_{unicast}^{mesh,\,RNDC} &= 2\lambda C^2 \\ BWeff_{unicast}^{mesh,\,RNDC} &= \frac{\sqrt{n}(\sqrt{n-1})}{\lambda C^2} = O\left(n^{-\frac{1}{3}}\right) \end{aligned} \tag{6.13}$$

Table 6.4 The complexity of different topology implementing Hopfield network

	NumLink	HopCount	ThroughPut	Performance-Cost
Mesh unicast	$O(n)$	$O(n\hat{}1.5)$	$O(n\hat{}-0.5)$	$O(n\hat{}-1.5)$
Mesh multicast	$O(n)$	$O(n)$	$O(1)$	$O(n\hat{}-1)$
Mesh broadcast	$O(n)$	$O(n)$	$O(1)$	$O(n\hat{}-1)$
Binary tree unicast	$O(n)$	$O(nlog_2 n)$	$O(\log_2 n\hat{}-1)$	$O(n\hat{}-1 \cdot (log_2 n)\hat{} - 1)$
Binary tree multicast	$O(n)$	$O(n)$	$O(1)$	$O(n\hat{}-1)$
Binary tree broadcast	$O(n)$	$O(n)$	$O(1)$	$O(n\hat{}-1)$
Fat tree unicast	$O(\text{nlog}_2 n)$	$O(\text{nlog}_2 n)$	$O(1)$	$O(n\hat{}-1 \cdot (log_2 n)\hat{} - 1)$
Fat tree multicast	$O(\text{nlog}_2 n)$	$O(n)$	$O(log_2 n)$	$O(n\hat{}-1)$
Fat tree broadcast	$O(\text{nlog}_2 n)$	$O(n)$	$O(log_2 n)$	$O(n\hat{}-1)$

Table 6.5 The complexity of different topology implementing RND network

	NumLink	HopCount	ThroughPut	Performance-Cost
Mesh unicast	$O(n)$	$O(\lambda C^2)$	$O(n^{-\frac{1}{3}})$	$O(n^{-\frac{4}{3}})$
Mesh multicast	$O(n)$	$O(\lambda C)$	$O(n^{\frac{1}{6}})$	$O(n^{-\frac{5}{6}})$
Mesh broadcast	$O(n)$	$O(n)$	$O(1)$	$O(n^{-1})$
Binary tree unicast	$O(n)$	$O(n^{\frac{1}{3}} logn)$	$O(n^{\frac{2}{3}} logn)$	$O(n^{-\frac{1}{3}} logn^{-1})$
Binary tree multicast	$O(n)$	$O(n^{\frac{1}{2}})$	$O(n^{\frac{1}{2}})$	$O(n^{-\frac{1}{2}})$
Binary tree broadcast	$O(n)$	$O(n)$	$O(1)$	$O(n^{-1})$
Fat tree unicast	$O(nlogn)$	$O(n^{\frac{1}{3}} logn)$	$O(n^{\frac{2}{3}})$	$O(n^{-\frac{1}{3}} logn^{-1})$
Fat tree multicast	$O(nlogn)$	$O(n^{\frac{1}{2}})$	$O(n^{\frac{1}{2}} logn)$	$O(n^{-\frac{1}{2}})$
Fat tree broadcast	$O(nlogn)$	$O(n)$	$O(nlogn)$	$O(n^{-}1)$

For the multicast, the total number of hops is approximated by the average distance when traversing one hop to the C connected ones:

$$\begin{aligned} TotalHopCount_{multicast}^{mesh,\,RNDC} &= (2\lambda + 1)\,C \\ BWeff_{multicast}^{mesh,\,RNDC} &= \frac{2\sqrt{n(\sqrt{n}-1)}}{(2\lambda+1)C} = O\left(n^{\frac{1}{6}}\right) \end{aligned} \tag{6.14}$$

Broadcast has the same hop count comparing to fully connected:

$$BWeff_{broadcast}^{mesh} = \frac{2\sqrt{n}}{\sqrt{n+1}} = O(1) \tag{6.15}$$

To simulate RNDC network in tree topology, the distance is calculated as:

$$D(n) = log_2 n \tag{6.16}$$

The average hop count for unicast is:

$$\begin{aligned} avgDist &= \frac{C}{2\pi\lambda^2} * \left(\frac{2}{e}\right)^{logn} \frac{logn-1}{2e^{-\frac{2}{\lambda}}-1} = O\left(n^{-\frac{1}{6}} logn\right) \\ TotalHopCount_{unicast}^{tree,\,RNDC} &= O\left(n^{\frac{1}{3}} logn\right) \\ BWeff_{unicast}^{tree,\,RNDC} &= \frac{2(n-1)}{TotalHC} = O\left(n^{\frac{2}{3}} logn^{-1}\right) \\ BWeff_{unicast}^{fatTree,\,RNDC} &= \frac{nlogn}{TotalHC} = O\left(n^{\frac{2}{3}}\right) \end{aligned} \tag{6.17}$$

The average hop count for multicast is first to average hop count of unicast and then to the C neuron in the branches:

$$\begin{aligned} TotalHopCount^{tree,\ RNDC}_{multicast} &= avgDist + C = O\left(n^{\frac{1}{2}}\right) \\ BW eff^{tree,\ RN\ DC}_{multicast} &= \frac{2(n-1)}{O\left(n^{\frac{1}{2}}\right)} = O\left(n^{\frac{1}{2}}\right) \\ BW eff^{fatTree,\ RN\ DC}_{multicast} &= \frac{nlogn}{O\left(n^{\frac{1}{2}}\right)} = O\left(n^{\frac{1}{2}} logn\right) \end{aligned} \quad (6.18)$$

The broadcast performance is the same as Hopfield connection.

The complexity of different topology implementing RND network is shown in Table 6.5.

6.4 Experimental Results

Throughput is measured by counting the packets arriving at each output. For uniform traffic, half of the traffic, N/2 packets, must cross the B_c bisection channels. The optimum throughput occurs when the packets are distributed evenly across the bisection channels. Capacity is defined as the bandwidth injected to each node for uniform random traffic, and can be calculated from the bisection bandwidth (B_B) or the bisection channel count (B_C) and the channel bandwidth (b) as [10]:

$$Capacity = 2B_B/N = 2bB_C/N \quad (6.19)$$

The ideal latency is dominated by the dimension. The average minimum distance hopcount H is [12]:

$$H = \begin{cases} \frac{2k}{3} & k\ is\ even, \\ \frac{2k}{3} - \frac{2}{3k} & k\ is\ odd \end{cases} \quad (6.20)$$

where k is the size of the array. The latency can be separated into router latency T_r (based on the stages of router structure) and interface latency T_i (depends on the techniques used for transmission). The ideal latency, which can also be regarded as zero-load latency, is $T_r * H + T_i * (H - 1)$.

The performance is tested with four types of traffic pattern, namely the random uniform traffic, the complete connection, the normal distribution traffic, and the neighbor connection. The uniform traffic is tested for the average performance, which occurs when the destination of packets are randomly generated. The complete connection traffic pattern is the worst case the network can face, i.e., every packet needs to be transferred to all other nodes; normal distribution of connection distance reflects the common case, and the neighbor connection condition is specified in [1].

The performance of the network is analyzed in the uniform traffic distribution. To find the maximum throughput, the offered traffic is changed under the configuration of four virtual channels per input port. The capacity of input traffic is one packet per node per cycle maximum, and the injection rate is 100%. As the injection rate increases, the throughput increases until the injection gets to 45%. The latency has a burst increment when the injection rate is greater than 45%, which means the network saturates. The number of virtual channels has strong influence on the total performance. The maximum throughput increases from 12 to 26 flits per cycle with the increment of virtual channels. When the number of virtual channel is greater than four, the improvement of performance is not obvious.

The performance is also tested with all-to-all connection pattern for the worst-case test (every simulated neuron in one FPGA generate data to all others). The whole transfer latency is recorded, which is the interval between the first packet being injected into the queue and the last packet being transferred to the destination node.

The total number of cycles required to send all packets of the test pattern from input to the output of each node is shown in Figure 6.5. By increasing the number of FPGAs, the total transfer cycles increase linearly from 513 to 4379 considering the interface latency, and from 351 to 2913, without interface latency. The average transmission latency of each packet is shown in Figure 6.6. When the dimension increases, there is a linear relation with the latency. For different dimension sizes, the average latency increases linearly from 209.03 to 2035.36, considering the interface latency, and from 125.94 to 1390.02, without interface latency. When the dimension increases, more contentions occur, and, consequently, the average throughput decreases. For no interface latency condition, the average node throughput keeps between 0.8 and 0.85 flits per cycle per node. When the interface latency is 4, the average throughput drops to around 0.6 to 0.55. The result shows that with the same router resource the interface latency decreases the throughput performance by 30% and increases the total transfer cycles by 46%.

Considering the interface latency in Figure 6.7, the throughput is decreased due to the idle cycle generated during the transferring. The test pattern is set to transfer all packets generated for the all-to-all connection.

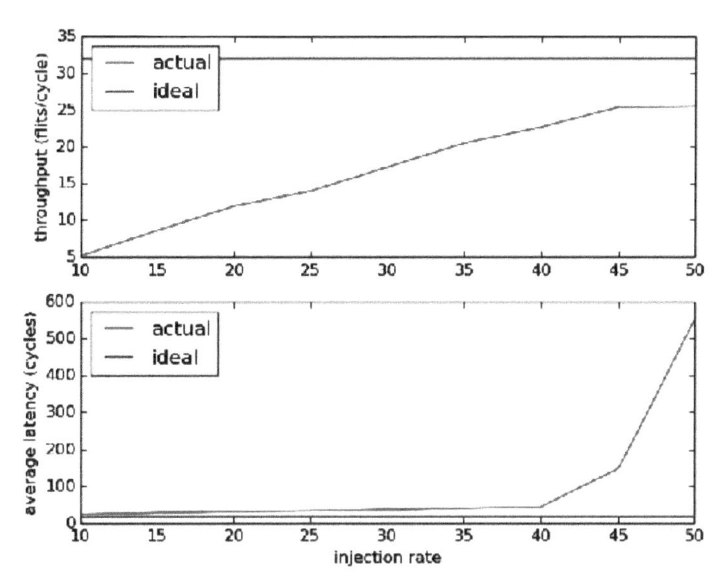

Figure 6.5 Throughput and latency with different injection rates (shown as units of percentage).

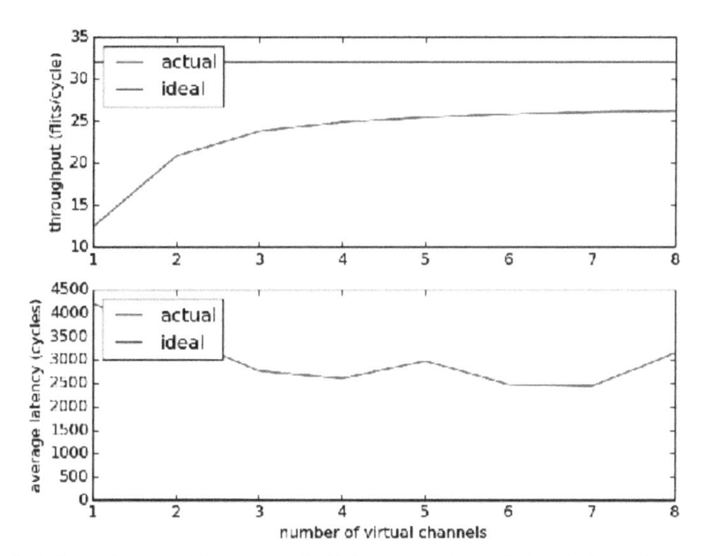

Figure 6.6 Throughput and latency with different number of virtual channels per input. To test the maximum throughput, the packets are generated with 100% input rate. The latency shows that the network is in the saturation condition.

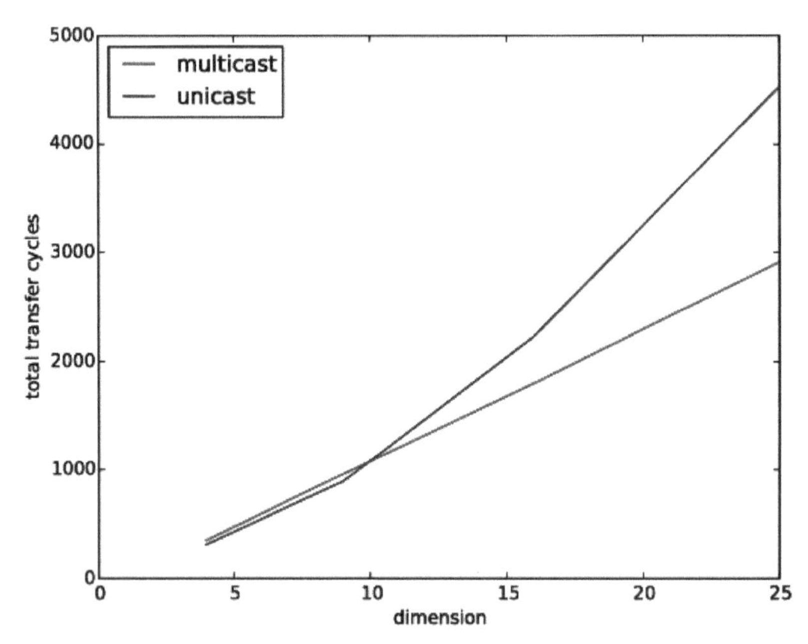

Figure 6.7 Total transfer cycles comparing with unicast and multicast for all-to-all connection with different mesh sizes. The network is configured with 100 packets per node.

The injection rate is set to 100, which means every cycle a packet is injected into the input queue.

The result in Figure 6.8 indicates that for the all-to-all connection traffic pattern, the mesh requires 15% less transfer cycles than other configurations. To meet the time constraint of 50 μs for each simulation step, the maximum number of neurons that can be simulated is approximately 5000 for the 100 MHz clock frequency. For the neighbor connected traffic pattern illustrated in Figure 6.9, the latency has linear relationship with the number of neurons per FPGA. Since the simulated neurons can be mapped to linear connection of FPGA, and one only communicates with the two neighbor FPGAs, no conflict occurs. In this specific condition, there is no limit on the number of cells that can be simulated. For normal distribution, the transfer cycles have a linear relationship with the number of simulated neurons. The number of neurons can reach 50k within the 5k cycles latency limit in comparison to 161 neurons in [4] and 400 in [13].

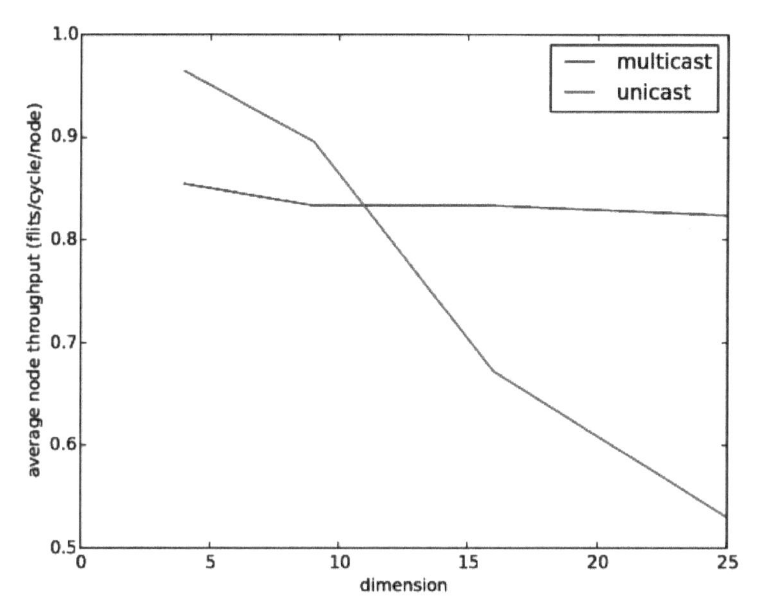

Figure 6.8 Average node throughput comparing with unicast and multicast for all-to-all connection with different mesh size. The network is configured with 100 packets per node.

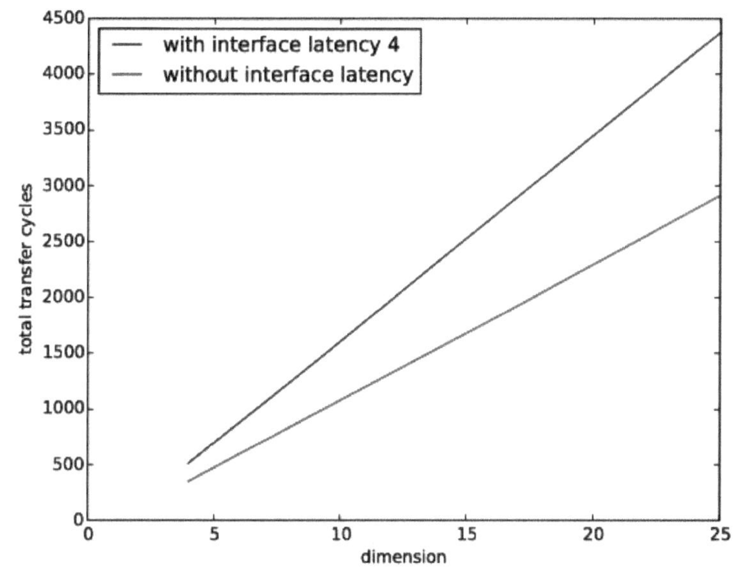

Figure 6.9 Simulation with interface latency. The network is configured with multicast communication. Each port has four virtual channels. The traffic pattern is fully connected with 100 packets sent per node.

6.5 Conclusions

Current neuron simulators, which are precise enough to simulate neurons in a biophysically meaningful way, are limited in amount of neurons to be placed on the chip, the interconnect between the neurons, run-time configurability, and the re-synthesis of the system. In this chapter, we propose a hierarchical dataflow architecture that is capable of bridging the gap between biophysical accuracy and large numbers (50k) of cells.

References

[1] G.J. Christiaanse, A. Zjajo, C. Galuzzi, R. van Leuken, "A real-time hybrid neuron network for highly parallel cognitive systems," International Conference of the IEEE Engineering in Medicine and Biology Society, pp. 792–795, 2016.

[2] M. van Eijk, C. Galuzzi, A. Zjajo, G. Smaragdos, C. Strydis, R. van Leuken, "ESL design of customizable real-time neuron networks," IEEE Biomedical Circuits and Systems Conference, pp. 671–674, 2014.

[3] J. Hofmann, A. Zjajo, C. Galuzzi, R. van Leuken, "Multi-chip dataflow architecture for massive scale biophysically accurate neuron simulation," International Conference of the IEEE Engineering in Medicine and Biology Society, pp. 5829–5832, 2016.

[4] H. Shayani, P.J. Bentley, A.M. Tyrrell, "Hardware implementation of a bio-plausible neuron model for evolution and growth of spiking neural networks on FPGA," NASA/ESA Conference on Adaptive Hardware and Systems, pp. 236–243, 2008.

[5] K.L. Rice, M.A. Bhuiyan, T.M. Taha, C.N. Vutsinas, M.C. Smith, "FPGA implementation of Izhikevich spiking neural networks for character recognition," International Conference on Reconfigurable Computing and FPGAs, pp. 451–456, 2009.

[6] D. Vainbrand, R. Ginosar, "Network-on-chip architectures for neural networks," ACM/IEEE International Symposium on Networks-on-Chip, pp. 135–144, 2010.

[7] J. Navaridas, M. Luan, J. Miguel-Alonso, L.A. Plana, S. Furber, "Understanding the interconnection network of Spinnaker," ACM International Conference on Supercomputing, pp. 286–295, 2009.

[8] S. Kumar, A. Jantsch, J.-P. Soininen, "A network on chip architecture and design methodology." VLSI, 2002. Proceedings. IEEE Computer Society Annual Symposium on. IEEE, 2002.

[9] J.R. de Gruijl, P. Bazzigaluppi, M. de Jeu, C.I. de Zeeuw, "Climbing fiber burst size and olivary sub-threshold oscillations in a network setting," PLoS Computational Biology, 8(12):e1002814, 2012.

[10] E. Bolotin, I. Cidon, R. Ginosar, A. Kolodny, "Cost considerations in network on chip," Integration – the VLSI Journal, vol. 38, no. 1, pp. 19–42, 2004.

[11] R. Legenstein, W. Maass, "Edge of chaos and prediction of computational performance for neural circuit models," Neural Networks, vol. 20, no. 3, pp. 323–334, 2007.

[12] N.E. Jerger, L.-S. Peh, "On-chip networks," Synthesis Lectures on Computer Architecture, vol. 4, no. 1, p. 141, 2009.

[13] M. Beuler, A. Tchaptchet, W. Bonath, S. Postnova, H.A. Braun, "Real-time simulations of synchronization in a conductance-based neuronal network with a digital FPGA hardware-core," International Conference on Artificial Neural Networks, pp. 97–104, 2012.

7

Single-Lead Neuromorphic ECG Classification System

Eralp Kolagasioglu

Delft University of Technology, Delft, The Netherlands

The pathophysiological processes underlying the ECG tracing demonstrate considerable variations in the morphological pattern/heart rate for different patients, or even in the same patient under different physical/temporal conditions. In this chapter, we propose a novel, energy-efficient, neuromorphic, global ECG classifier applicable for an unsupervised learning, which offers an effective platform to provide personalized prognosis by combining the heterogeneous sources of available information and identifying meaningful patterns in data. The implemented system utilizes adaptive ECG interval extraction for feature extraction, and correlation matrix for unsupervised feature selection. The resulting features are encoded into spikes and offered to neuromorphic liquid-state machine spiking neural network for classification. For each clustering, the Silhouette coefficients have been calculated, both based on the selected features and the spike times of the pool for each heartbeat. For compatibility with the wearable devices, the classification system requires only one ECG lead. The experimental results indicate that the proposed design can accurately classify seven heart beat types with an overall classification accuracy of 95.5% at the cost of less than 1 μW.

7.1 Introduction

This section offers a brief introduction to ECG signals and the classification process for ECG signals. The elaborations will focus on explaining how the system works, giving examples of the state-of-the-art solutions and the restrictions set by the problem.

7.1.1 ECG Signals and Arrhythmia

Electrocardiography (ECG) measures the electrical activity of the heart using the potential difference between different leads placed on the body. A typical healthy ECG heartbeat can be seen in Figure 7.1. An ECG beat consists of four typical complexes: P, QRS, T, and U. Since the U wave cannot be distinguished most of the time, it is not illustrated. The P wave indicates the contraction of the atrial rooms of the heart, the QRS complex shows the contraction of the ventricular rooms, and finally the T wave is the relaxation of the ventricular rooms. The shape and size of these waves vary based on the lead that the signal is being observed from. As a result, different features can be seen from different leads, which makes diagnosing particular diseases easier from certain leads. However, most of the arrhythmias can be distinguished from the modified lead two (ML-II lead), which can be seen in Figure 7.2. Moreover, the aim of this chapter is to illustrate a system that can potentially be implemented as a wearable device, which limits the system requirements to only a three-lead ECG instead of a hospital level 12-lead ECG. Figure 7.1 illustrates some of the typical features of the ECG signal that are examined to detect abnormalities. Arrhythmias designate abnormalities in the rhythm of the heart (i.e., heart rate). An example of arrhythmia can be seen in Figure 7.3. By examining the features other than the heart rate, differentiation between a regular and a sick beat can be made, in addition to

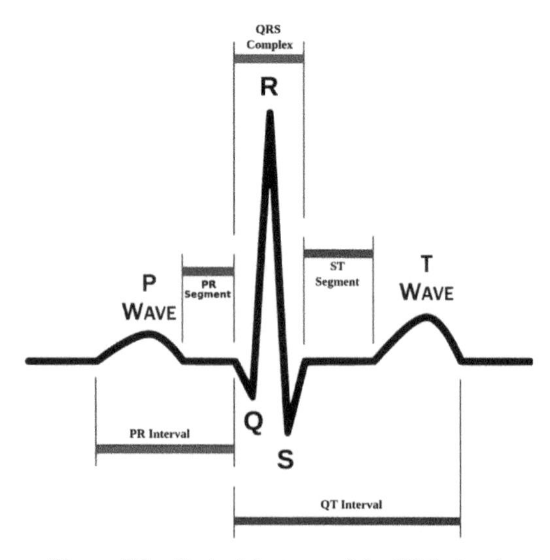

Figure 7.1 Typical features of the ECG signal.

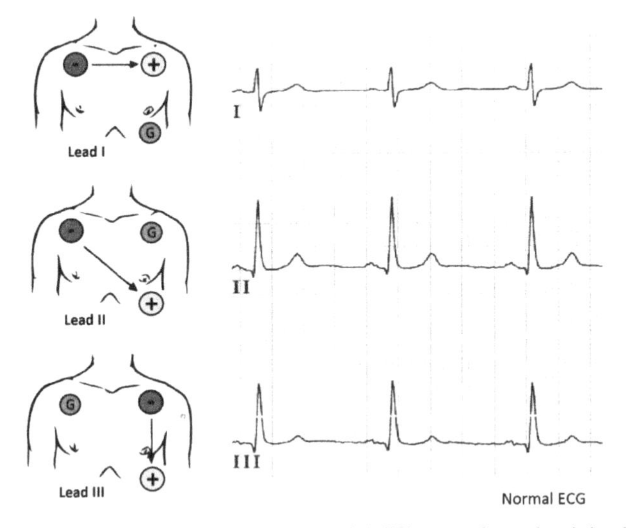

Figure 7.2 Placement of 3-lead ECG: the potential difference from the right shoulder to the left hip designates the ML-II signal.

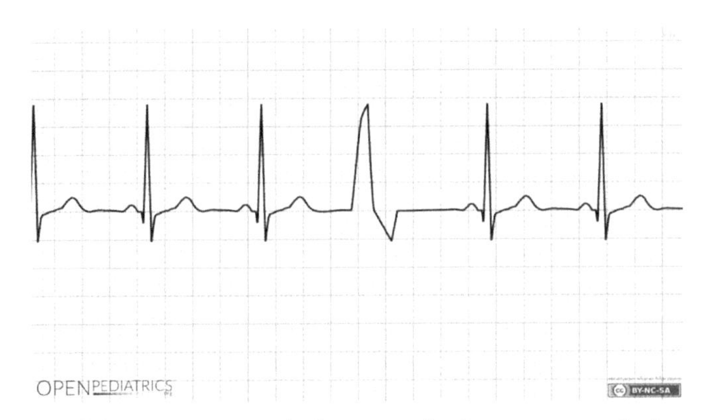

Figure 7.3 Premature ventricular contraction beat among regular beats.

assisting to differentiate between different beat types. The beat types found in the MIT-BIH Arrhythmia database [1] can be seen in Table 7.1.

All the features have a pre-determined normal range in an ECG signal. However, in general, the inter-patient variations of the ECG signals and the stochastic nature of the main pathophysiological processes result in the inconsistent performance, and consequently, to significant variations in their accuracy and efficiency. In this context, unsupervised machine learning may offer an effective platform. This has been strongly emphasized in [2].

Table 7.1 Arrhythmia types and their respective annotations in MIT-BIH database

Annotation	Meaning
N	Normal beat
L	Left bundle branch block beat
R	Right bundle branch block beat
A	Atrial premature beat
a	Aberrated atrial premature beat
J	Nodal (junctional) premature beat
S	Supraventricular premature beat
V	Premature ventricular contraction
F	Fusion of ventricular and normal beat
!	Ventricular flutter wave
e	Atrial escape beat
j	Nodal (junctional) escape beat
E	Ventricular escape beat
/	Paced beat
f	Fusion of paced and normal beat
Q	Unclassifiable beat
|	Isolated QRS-like artifact

7.1.2 Feature Detection

Feature detection in ECG signals primarily focuses on the signal from the ML-II lead. The main property of this signal is (as can be seen from Figure 7.1) that the R waves are quite large and dominant. Thus, the algorithms mainly focus on first detecting the QRS complex, and subsequently on finding the P and the T waves. In majority of the algorithms, the ECG signal is first preprocessed in the bandpass filter to suppress noise and to accentuate the QRS complex. The resulting signal then goes through an adaptive thresholding phase to find the QRS complexes. Finally, if the algorithm supports it, the P and T waves are searched for in small windows around the QRS complex. This section aims at exploring some of the possible algorithms for feature detection and comparing them with respect to the goal of this chapter.

7.1.2.1 Methods and algorithms

7.1.2.1.1 *QRS detection*

Majority of the previous art focus on the QRS detection of ECG signals, since they are much easier to locate due to the high-amplitude R spike in the ML-II lead of an ECG (Figure 7.1). Usually, the algorithms depend on preprocessing of the signal to obtain a cleaner signal with less noise, transforming the

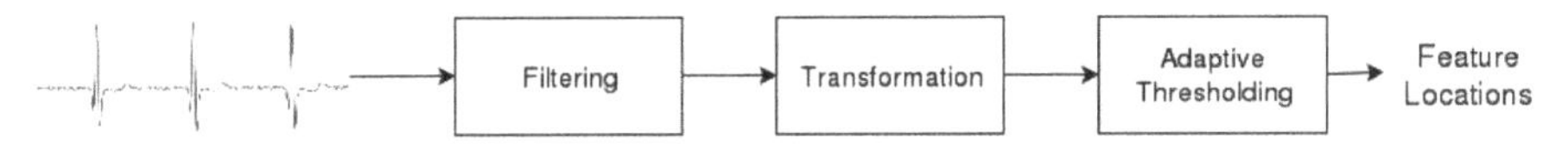

Figure 7.4 Flow diagram for feature detection.

signal in a way that would emphasize the QRS complex, and finally using an adaptive thresholding method to find the QRS complex locations. A flow diagram for this process can be seen in Figure 7.4.

An exception is [3], which uses the energy of the signal to locate the QRS complex (making use of the high amplitude R peak). In [5], a curve length transform is used to calculate the length of a line within a given window. Since the QRS complex has a spike, the window that contains the spike has a much higher value than the others: this property is subsequently exploited to locate beats.

In [4], after initial bandpass filtering, the signal is passed through a derivative filter to accentuate the QRS complex since it has a high slope spike in it. The output of this filter is used to find the onset and end points as well, such as in [7]. At the end, the signal is squared to remove the negative parts and finally passed through moving-window integration.

Wavelet transforms are used in [6]. Different transform levels of an ECG can be seen in Figure 7.5. The peaks are called the modulus maxima points of a signal. Distinct shapes are obtained in the wavelet domain, for each different shape in the time domain. An adaptive thresholding algorithm has been used to find the possible QRS complexes in the wavelet domain.

In Table 7.2 a comparison of the implemented algorithms [4], [5], and [6] can be seen in terms of complexity. In addition, their accuracy statistics on the MIT-BIH database can be seen in Table 7.3.

7.1.2.1.2 *P and T wave detection*

P and T wave detection is a more arduous task compared to QRS complex detection mostly due to their low amplitude. In addition, it is also more difficult to find the onset and end points due to their much smaller slope in comparison with the QRS complex. Based on [8], [6] performs better than [7]. In [7], [4] is used for the QRS detection and then the output of the derivative filter from [4] again for finding P and T waves. In contrast, [6] utilizes the wavelet transform for detection. Finally, [8] uses a multiscale morphological derivative for detection of wave boundaries.

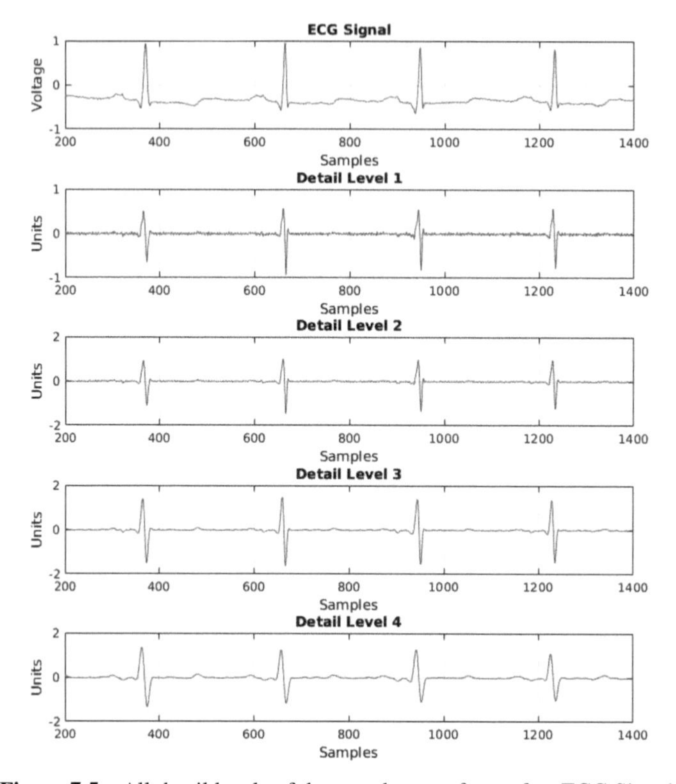

Figure 7.5 All detail levels of the wavelet transform of an ECG Signal.

7.1.3 Feature Selection

Having the onset, end and peak points of P, QRS, and T complexes, multitude of the signal features can be obtained. However, not all of these features would be necessary to detect the beat types we are interested in. Furthermore, most diseases have very few indicators and using too many features can lead to their approximation to healthy beats, i.e., the fact that they have only one feature that is different than healthy beats and dozen features that are the same as healthy beats might lead to misclassification. In addition, using fewer features lowers training time and network size, as well as allowing lower power consumption. The literature covering this topic is limited for ECG signals, except for [2]; existing classification methods in Section 7.1.4 focus on fewer disease groups, which allows them to use a much lower and select feature count.

Table 7.2 Complexity comparison of QRS detection algorithms (L is the length of the signal). "Transforms" show the number of operations required for each algorithm, to transform the signal to their respective domain. "Learning" indicates the required number of steps for [5] to learn a threshold. Finally, "Detection" shows the number of operations required in each iteration (sample) during the detection phase. "If Not Beat" is used to indicate the number of operations required if the sample being examined is detected as non-beat sample and conversely for "If Beat". "If Beat" requires more operations since it requires finding the onset and end points

	[5]	[4]	[6]
	Transforms	Transforms	Transforms
Add	61L	67L	16L
Mul/Div	7L+1	71L	24L
Square	L	L	–
Square root	L	–	–
For each sample	Learning	Learning	Learning
Add	4	–	–
Min	2	–	–
Compare	5	–	–
Mul/Div	2	–	–
For each sample	Detection	Detection	Detection
	If not Beat	If not Beat	If not Beat
Add	1	–	4
Min	2	–	1
Compare	1	3	7
Abs	–	–	2
RMS over	–	–	720
	If Beat	If Beat	If Beat
Add	622	615	799
Min	–	7	9
Compare	629	631	1009
Mul/Div	28	18	5
Abs	–	–	319
Sort	–	1	3
RMS over	–	–	54000

Table 7.3 Average accuracy over all patients in MIT-BIH on finding beats for the implementation of all three algorithms

Evaluation	[5]	[4]	[6]
Average beat count	2289	2267	2266
True positives	2261	2247	2215
False positives	28	20	51
False negatives	23	37	69
True positive rate	98.98%	98.48%	96.92%
Positive predictive value	98.73%	99.13%	97.58%

7.1.3.1 Feature selection choices

The different options for feature selection can be seen in Figure 7.6. The first option is to whether to use a supervised or an unsupervised feature selection algorithm. Supervised feature selection algorithms are embedded, and wrapper methods as those are the ones that use the classifier as part of the feature selection process. Most supervised algorithms use a greedy approach (although expensive) to find the feature set that offers the best classification results with the given classifier. However, due to interpatient variation, labeled heartbeats might not be a realistic case.

7.1.3.2 Methods and algorithms

There are three main methods for feature selection: filter, embedded, and wrapper methods [9]. Filter methods rely on the intrinsic statistical properties of the given feature set, for example, by examining the correlations within the data using a correlation matrix or creating a new orthogonal basis for the feature set with principal component analysis (PCA) or independent component analysis (ICA) to create a reduced feature set. Embedded and wrapper methods, on the other hand, use the output of the learning machine (classifier) and optimize the output with various feature sets. As a result, embedded and wrapper methods are usually more accurate, yet computationally more expensive. The main difference between these two methods is that embedded methods have an added metric or penalty in the learning process.

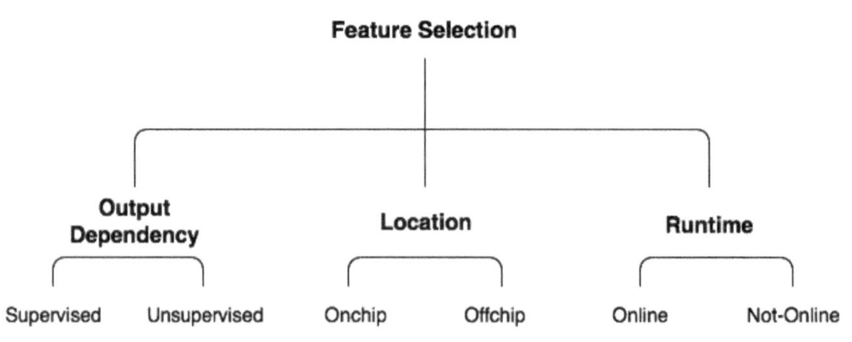

Figure 7.6 Choices to be made for feature selection.

Table 7.4 State-of-the-Art comparison

	Acc.	Learning	Features	Diseases	Power	Tech.	Area	Supply
[10]	98.41%	Supervised	NA	4	NA	NA	NA	NA
[11]	98.9%	Supervised	18	2	NA	NA	NA	NA
[12]	86%	Supervised	10	Prediction	2.78 μW	65 nm	0.112 mm	1V
[13]	∼99%	Unsupervised	3+Beat	2	NA	NA	NA	NA
[14]	NA	Supervised	7+Beat	5 Classes	NA	NA	NA	NA
[15]	98.5%	Unsupervised	12	16	NA	NA	NA	NA
[16]	97.25%	NA	Beat	3	5.967 μW	0.18 μW	2.465 mm	1.2V
[17]	87%	Supervised	3	4	112.7 μW	65 nm	NA	NA
[18]	99%	Unsupervised	9	2	NA	NA	NA	NA
[19]	97%	Unsupervised	5	3	NA	NA	NA	NA
[20]	97%	Supervised	6	4	NA	NA	NA	NA

7.1.4 Classification Methods

A comparison of various methods can be seen in Table 7.4. Only [15] is a global classifier, with all the other classifiers focused on a small number of beat types.

7.2 Feature Extraction Implementation

A block diagram showing the overview of feature extraction can be seen in Figure 7.7.

7.2.1 Feature Detection

7.2.1.1 QRS detection

Preprocessing: During preprocessing, the signal is passed through an FIR bandpass filter to suppress the muscle noise, 60 Hz interference from the

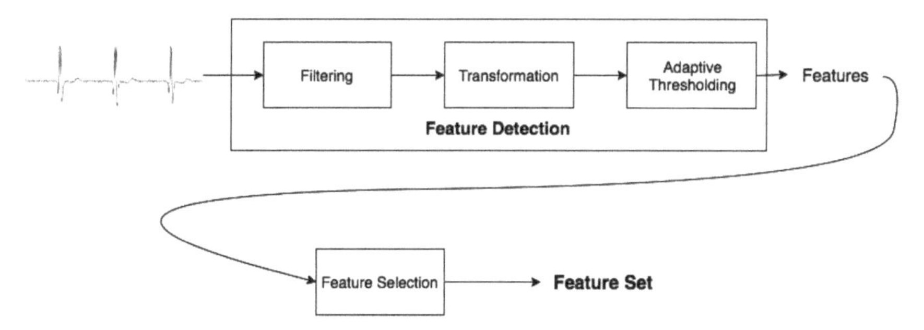

Figure 7.7 Feature extraction block diagram.

power line, and baseline wander present in ECG signals. The filter is designed for a sampling frequency of 360 Hz, and has a stop band attenuation of 40 dB.

Signal Transformation: Curve length transform for discrete signals that calculates the length of a line within a given window is described as [5].

$$L\left(w,i\right)=\sum_{k=i-w}^{i}\sqrt{\Delta t^{2}+\Delta y_{k}^{2}} \tag{7.1}$$

where w is the window length, i is the index, Δt shows the sampling period, and Δyk = yk – yk –1. The shape of the signal after transformation can be seen in Figure 7.8. The window size is suggested to be as wide as the widest QRS complex. If the window size selected is smaller than a QRS complex, that QRS complex will create multiple peaks after the transformation. If the converse is done, a very large window will result in the subsequent T wave to be involved in the window, and the transformation will reach its maximum point later than it is supposed to. As a result, the end point of the QRS could be misdetermined. The relationship between the signal after the transformation and the actual QRS complex can be seen in Figure 7.9.

QRS Detection: After this transformation, an adaptive thresholding algorithm searches for the QRS complexes. After it finds a complex, it looks for the points at which the transform meets the baseline and the point at which it

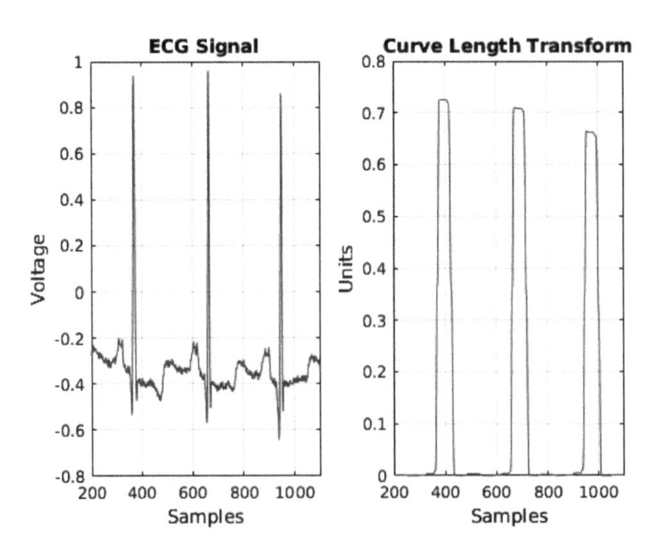

Figure 7.8 ECG signal and its curve length transform.

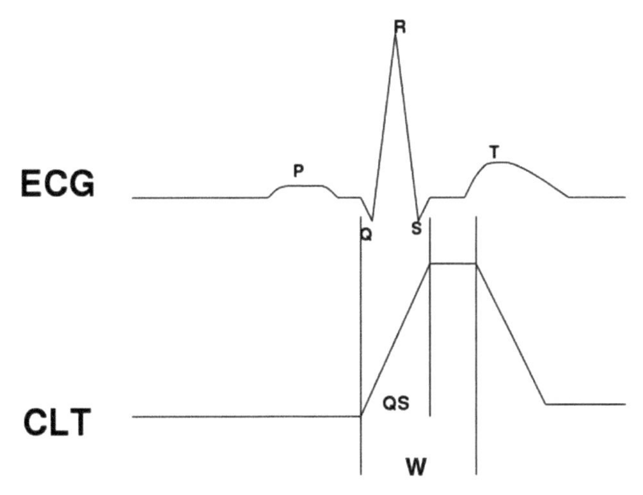

Figure 7.9 Onset and end points representation on CLT.

reaches its maximum point. The pseudocode for the training can be seen in Algorithm 1, and the QRS detection algorithm can be seen in Algorithm 2.

The values in the algorithm that are used to adjust the thresholds are based on the amplitude of the curve length transformed signal. The eye-Closing parameter has been set for power-saving concerns. A successful and unsuccessful onset and end detections are illustrated in Figures 7.10 and 7.11, respectively.

7.2.1.2 P and T wave detection

After the QRS peaks detection, the P and T waves are only searched in small windows before and after the QRS. Thus, the wavelet transform is only applied to those small windows around the QRS complex instead of the whole signal. Window widths have been determined empirically. For the P wave, the window size is (heartrate)/4, where heartrate is in samples and the maximum window size is 410 ms (150 samples). For the T wave, on the other hand, the window size is (heartrate)/1.9, where heartrate is in samples and the maximum window size is 555 ms (200 samples).

The complexity of the algorithm is much reduced when QRS complexes are already located and only the P and T waves are searched for. The algorithm searches for the peaks can be seen in Figure 7.5; these peaks are called the modulus maxima points of a signal. While for QRS detection, the algorithm goes through all the detail levels of the signal, for P and T wave detection, only the fourth level is searched. For higher accuracy, if

Algorithm 1 Training for Initial Threshold Calculation

```
1: Parameters :
2: sfreq ← sampling frequency
3: learnTime ← 8 * sfreq
4: adjustThreshold ← 10
5: thresholdMaxDiv ← 3
6: eyeClosing ← round (sfreq * 0.25)
7: QRSheightThresh ← 0.002
8: minThresh ← 0.01
9: notFoundStep ← 0.01
10: timer ← 0
11: Texpected ← sfreq * 2.5
12: flatRange ← 25
13: QRSoffset ← 5
14: InitialValues :
15: thresh ← 0
16: i ← 1
17: Tlower ← Average of transformSignal
18: Tupper ← Tlower * thresholdMaxDiv
19: procedure INITIAL TRAINING
20:     while i < learnTime do
21:         thresh ← 2 * Tlower
22:         if transformSignal(i) > thrash then
23:             Lmax ← Maximum value of transformSignal from i to i + eyeClosing/2
24:             Lmin ← Maximum value of transformSignal from i − eyeClosing/2 to i
25:             if Lmax > Lmin + QRSheightThresh then
26:                 Tupper ← Tupper + (Lmax − Tupper)/adjustThrehsold
27:                 thresh ← Tupper/thresholdMaxDiv
28:     i ← i + 1
29:     thresh ← Tlower
```

it cannot be found on the fourth level, the next one can also be searched. One of the reasons to start looking from a higher level is that larger scales have fewer modulus maxima points, which improves the calculation time. In addition, higher scales greatly suppress the high-frequency noise components and prevent them from creating modulus maxima points. After finding the onsets and end points, the peak is found at the zero-crossing point of a positive maximum-negative minimum modulus maxima pair. The pseudocode for this algorithm can be seen in Algorithm 3. The same algorithm can be used for each wave.

Algorithm 3 consists of three procedures. The wavelet transform is modified in such a way that the maximum positive maximum modulus peak has

Algorithm 2 QRS Detection Algorithm

```
1:  procedure QRS DETECTION
2:      while i < Signal Length do
3:          if transformSignal(i) > thresh then
4:              Lmax ← Maximum value of transformSignal from i to i + eyeClosing/2
5:              Lmin ← Minimum value of transformSignal from i − eyeClosing/2 to i
6:              timer ← 0
7:              if Lmax > Lmin + QRSheightThresh then
8:                  Lonset ← Lmax/100
9:                  loop with j from i − eyeClosing/2 to i
10:                     if transformSignal(j to j+ flatRange) < Lonset then
11:                         QRSonsetPoint ← j − QRSoffset
12:                 loop with j from i to i + eyeClosing/2
13:                     if transformSignal(j) > Lmax − (Lmax − Lmin)/25 then
14:                         QRSend ← j + QRSoffset
15:                 i ← i + eyeClosing
16:                 Tupper ← Tupper + (Lmax − Tupper)/adjustThrehsold
17:                 thresh ← Tupper/thresholdMaxDiv
18:             timer ← timer + 1
19:             if timer > Texpected and Tupper > minThresh then
20:                 Tupper ← Tupper − notFaundStep
21:                 thresh ← Tupper/thresholdMaxDiv
22:         i ← i + 1
```

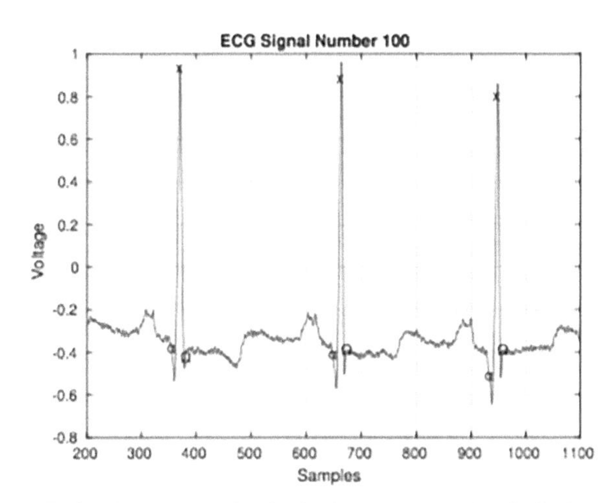

Figure 7.10 Good example of a QRS onset (α) and end (Ω) detection.

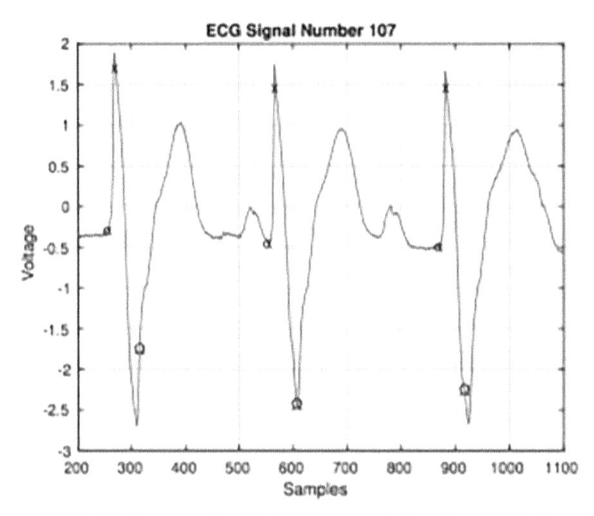

Figure 7.11 Bad example of a QRS onset (α) and end (Ω) detection.

Algorithm 3 Locating P or T waves using discrete wavelet transform, PMM is positive modulus maximum point NMM is negative modulus maximum point

---

1: **Parameters :**
2: *maxPiitindow* $\leftarrow$ 417*ms*
3: *Pwindow* $\leftarrow$ (*QRSonset* $-$ *min*(*HeartRate*/4, *maxPwindow*) to *QPSonset*)
4: *maxPwave* $\leftarrow$ 153*ms*
5: *PonsetThresh* $\leftarrow$ 0.5
6: *PendThresh* $\leftarrow$ 0.9
7: **Variables :**
8: *dwtSignal* $\leftarrow$ Discrete Wavelet Transform of Signal
9: *PsearchWindow* $\leftarrow$ *dwtSignal*(*Pwindow*)
10: **procedure** MODIFY PSEARCHWINDOW
11: *wt* $\leftarrow$ *PsearchWindow*
12: *PsearchWindow* $=$ *wt* - (*max*(*wt*) $-$ (*max*(*wt*) $-$ *min*(*wt*)/2)
13: *mm* $\leftarrow$ Modulus maxima points in *PsearchWindow*
14: **procedure** REMOVE NON-POSSIBLE ZERO-CROSSINGS
15: **if** $((mm(i+1)mm(i) < maxPwave)\&\&(mm(i+2) - mm(i+1) < maxPwave))\&\&sign(mm(i+1)) == +$ **then** mm(i) = Deleted
16: **if** $mm(i+1) - mm(i) > maxPwave$ **then** mm(i) = Deleted
17: **procedure** DETECT ONSET, END AND PEAK
18: *peak* $\leftarrow$ zero-crossing between the modulus maxima
19: *start* $\leftarrow$ point before PMM $<$ *PonsetThresh* $*$ *PMM*
20: *end* $\leftarrow$ point after NMM $<$ *PendThresh* $*$ *NMM*

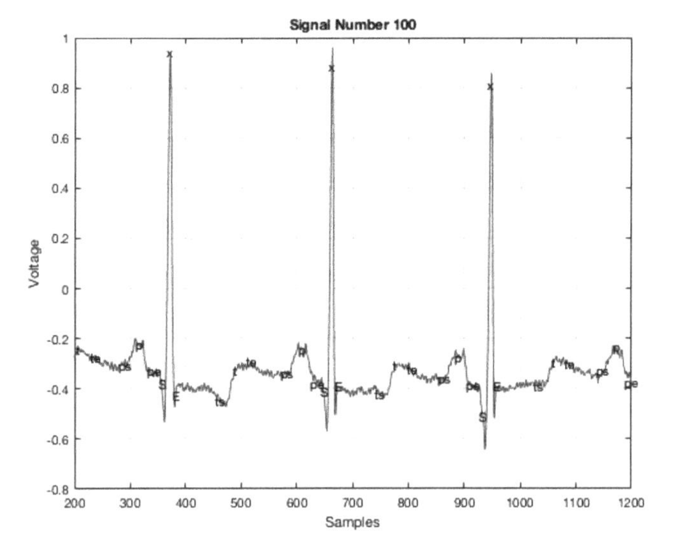

Figure 7.12 Annotations on the ECG Signal.

the same amplitude as the minimum-negative maximum modulus peak. This assured that both peaks that represent a signal are above the noise level for the same noise multiplier.

The next procedure in the algorithm removes the modulus maxima points, which do not represent a P or T wave. The first if condition eliminates a particular morphology trait, while allowing the wave to be still detected. The second if condition removes a point if it is too far from another modulus maxima point (so it cannot actually be representing a wave). Finally, the third procedure is for finding the onset, end, and peaks of the waves. For the T wave, the onset and the end thresholds are 0.05 and 0.4, respectively [6]. An illustration of the annotations of the algorithm can be seen in Figure 7.12.

7.2.2 Feature Selection

7.2.2.1 Feature set

The complete feature set includes:

- QRS Width: The time from the onset of the QRS complex to the end of the QRS complex
- R Voltage: The voltage level of the R peak
- QR Width: The time from the onset of the QRS complex to the R peak
- RS Width: The time from the R peak to the end of the QRS complex

- P Exists: Binary value showing whether the P wave exists
- PR Interval: The time from the onset of the P wave to the onset of the QRS complex
- PR Segment: The time between the end of the P wave to the onset of the QRS complex
- P Voltage: The voltage level of the P peak
- P Width: The time between the onset and the end of the P wave
- Heart Rate: The time between the R peaks of the current and the previous beat
- T Exists: Binary value showing whether the T wave exists
- PT Interval: The time between the onset of the P wave and the end of the T wave
- QT Interval: The time between the onset of the QRS complex and the end of the T wave
- ST Segment: The time between the end of the QRS complex and the onset of the T wave
- ST Interval: The time between the end of the QRS complex and the end of the T wave
- T Voltage: The voltage level of the T peak
- T Width: The time between the onset and the end of the T wave
- Previous Heart Rate: The time between the preceding beats R peaks
- PP: The time between the P peaks of the current and the previous beat
- Previous PP: The time between the preceding beats P peaks
- QRS Lowest Voltage: The lowest voltage level of the QRS complex

For scaling issues and to tackle outliners, a two-stage algorithm was developed. First, the outliers were determined and set to an average value, and in the next step, all the zeros were set to the min value within the observation set of those features. The pseudocode for this algorithm can be seen in Algorithm 4.

7.2.2.2 Correlation matrix

Pseudo code for this algorithm can be seen in Algorithm 5. The elimination of the features is based on three rules: (i) the features are designated as correlated if they have >80% correlation; (ii) due to their reliability, the correlated features associated with the QRS complex are kept; and (iii) the feature that is correlated to most features is kept. An example of a correlation matrix can be seen in Figure 7.13, which shows the correlation matrix as a heat map; the features that are highly correlated have a red hue, whereas the ones that have no correlation are white, and the ones that are negatively correlated are blue.

Algorithm 4 Handling the Outliers in the Observations

1: **Parameters :**
2: *observations* $\leftarrow$ all observations of a feature
3: *thresh* $\leftarrow$ 0.8
4: **procedure** FIXING THE OUTLIERS
5: $outliers \leftarrow observations > (1 + thresh) * mean(observations)$ && $observations <$ $(1\ thresh) * mean(0observations)$
6: $outliers \leftarrow mean(observations)$
7: **procedure** FIXING THE MISDETECTIONS
8: *misdetections* $\leftarrow$ All the zeros in the detection
9: $misdetections \leftarrow min(observations)$

Algorithm 5 Computing the Correlation Matrix

1: **Parameters :**
2: *features* $\leftarrow$ matrix of all observations of all features
3: **procedure** CORRELATION MATRIX CALCULATION
4: *features* $\leftarrow$ Standardize Features
5: $u \leftarrow$ mean value of each feature through all observations
6: $B \leftarrow$ deviations of each observation from the mean
7: C (Covarience Matrix) = $\frac{1}{n-1} B^* \otimes B$
8: Remove one of the features that are correlated more than 80% or 0.8

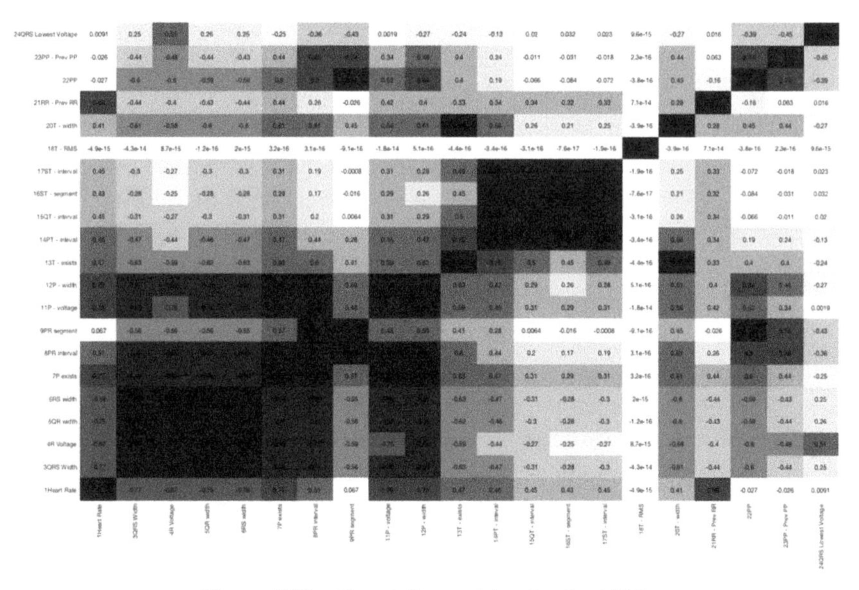

Figure 7.13 Correlation matrix of patient 100.

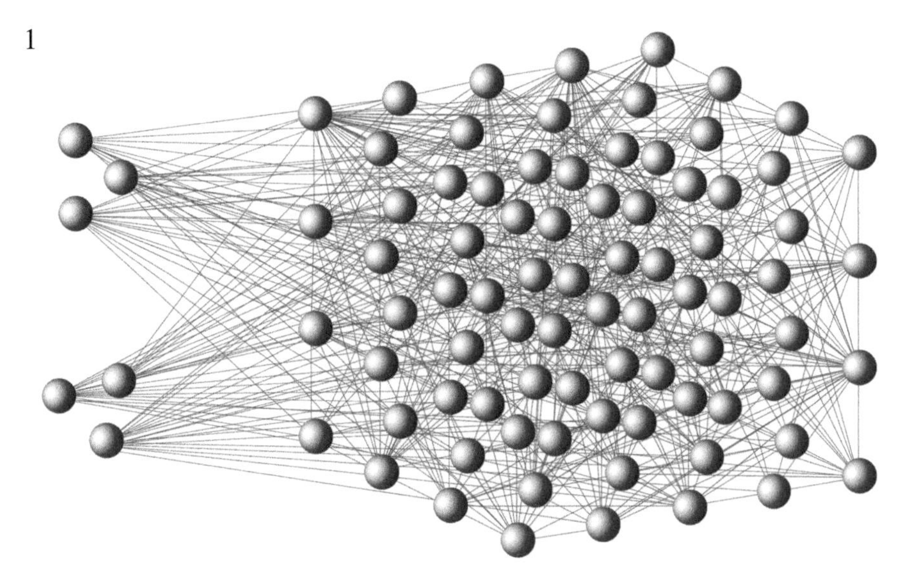

Figure 7.14 Liquid-state machine architecture.

Figure 7.14 shows the correlation matrix as a heat map, the features that are highly correlated have a red hue, whereas the ones that have no correlation are white and the ones that are negatively correlated are blue.

7.3 Network Configuration and Results

7.3.1 Approach

The ECG classification has been performed with liquid-state machine (LSM) [21] neuromorphic network. The LSM architecture mainly uses two layers of neurons. The first layer consists of the input neurons, and the second layer is a pool of neurons that are randomly connected among each other. As a result of this connection pattern, each distinct input pattern results in a different order of firing at the LSM pool. An illustration of an LSM network can be seen in Figure 7.14. Consequently, the inputs are coded in the time to spike values of the pool neurons.

The neurons in the population exhibit overlapping sensitivity profiles. The stimulation profile is determined by a Gaussian activation function; subsequently, this function determines each neurons receptive fields, i.e., each input stream is normalized, and based on the number of input neurons assigned for each input, receptive fields are set. The first step in the most appropriate and generalizable approach to LSM is to make the connections between the first and second layers more sparse, which provides the opportunity for different input patterns to trigger different parts of the pool first and increase the number of clusters.

The next step is to add a randomized delay matrix to the connections, which ensures that the time difference to spike between neurons increases and prevents neurons from firing if they have discharged. This reduces the classification complexity as the distinction between the patterns has increased. The final adjustment to the configurations is to lower the weight distributions, both from the first to second layer and the ones within the pool. For correct operation of the network, membrane voltage of each neuron needs to reach its resting value before feeding each input pattern, and the depression in the network should be strong enough such that a stable state can be reached that is not an all-pass network.

To find an optimal initial configuration, random initialization has been used. Alternatively, a semi-supervised approach could be used for the network training for some distinctive disease characteristics (e.g., two different beat types are almost identical to each other, a very low beat type count), or a supervised approach (requiring labeled data), such as a genetic algorithm based on [28], can be utilized. In general, it is observed that the lower limit of the weight range needs to be sufficiently high to facilitate the stimulation of an adequate number of neurons, yet prevent the activation of connections that are never re-used in the future. Similarly, the upper limit of the range should be sufficiently low to prevent over-training of synapses on the early training data.

To compensate for interpatient variations in the morphological pattern/heart rate, the classifier is adjusted and trained for individual characteristics of each patient. A pre-defined section of the recorded beats, which includes all patients beat types are offered to the classifier at least once. Triplet spike time-dependent plasticity [29] rule has been used for adjusting the weights among the neurons during the training phase. A resolution of five neurons is utilized for encoding of each feature, and the encoded spikes are fed to a pool of 25 neurons. Note that the accuracy of the system is proportional to the number of neurons, in either the pool or the inputs, at the cost of increase power consumption and complexity of the system, e.g., initial condition settings, training, and learning process.

7.3.2 Silhouette Coefficients

To evaluate the created clusters, i.e., to determine whether more clusters are required, whether the clustering methods are resulting in strong clusters, and whether the problem is with the network, or the features not sufficient for clustering, we employed Silhouette coefficients [22], which can be defined as:

$$s(i) = \frac{b(i) - a(i)}{\max(a(i), b(i))} \tag{7.2}$$

where $a(i)$ is the average distance of beat *i* to all the other beats in its cluster and $b(i)$ is the minimum of the average distances of beat *i* to all the other beats in other clusters. Thus, the smaller the $a(i)$ value, the better as it would show similarity of the beat to its own cluster, and the larger the $b(i)$ value, the better as it would show the dissimilarity to other clusters. It can be seen from (3.1) that:

$$-1 < s(i) < 1 \tag{7.3}$$

A value close to 1 shows that a beat belongs to its cluster, whereas a value close to -1 shows that the beat fits some other cluster more, and a value close to 0 shows that the beat fits into at least two clusters well, so it is right on the edge.

In the algorithm, any means can be used to calculate the distance between the elements of a cluster. We employed Manhattan distance; for each experiment, the Silhouette coefficients have been calculated, both based on the selected features and the spike times of the pool for each beat.

7.3.3 Clustering Methods for the Output

Several methods to cluster distinct output patterns of the spiking neural network exist, such as methods based on mean difference clustering, Otsu clustering [23], K-means clustering [24], minimum difference clustering, and sequential clustering. We implemented sequential clustering due to its consistent results and its online utilization. The pseudocode for this algorithm can be seen in Algorithm 6. The sequential algorithm uses the sum of absolute spike timing differences for every output neuron. In this algorithm, this value is calculated between the new coming beat and the average of each cluster. Next, these differences are compared to the set limit, which decreases by a percentage at each iteration. Finally, the beat is inserted in one of the existing clusters if it is deemed similar enough; otherwise, a new cluster is created for the beat. Based on the silhouette coefficients, this algorithm resulted in the strongest clusters.

Algorithm 6 Sequential Clustering for Spike Times

1: **Parameters :**
2: *seqDiff* $\leftarrow$ The initial difference between values to seperate clusters
3: *seqltPercentage* $\leftarrow$ The percentage by which the threshold is lowered every iteration
4: *iterationCount* $\leftarrow$ number of iterations
5: **Inputs :**
6: *spikeTimes* $\leftarrow$ Spike times for a beat
7: *beatCount* $\leftarrow$ Number of beats
8: **procedure** CLUSTERING
9: Put the first beat in a cluster
10: *currentClusterCount* $\leftarrow$ **1**
11: **for** $i = 1$ to *iterationCount* **do**
12: **for** $j = 2$ to *beatCourd* **do**
13: **for** $c = 1$ to *currentClusLerCoimt* **do**
14: $diffs = sum(abs(spikeTimes(j) - mean(spikeTimes(Clusters(c)))))$
15: *possible* $\leftarrow$ Clusters that have $diffs < (seqDiff * seqltPercentage^{i-1})$
16: Remove beat j from the clusters if it is already there
17: **if** *possible* is empty **then**
18: Create new cluster for beat j
19: **else**
20: Add to the cluster that has the least difference

7.3.4 Results

Table 7.5 illustrates the detection rate and the percentage of missed beats from the database due to detection. In terms of classification, the results of the clustering of detected beats can be seen in Table 7.6. The clustering of each beat has been evaluated on the basis of whether they are the dominant beat type in a cluster. The proposed solution is designed with the constraint of being able to be adapted to a wearable device, and subsequently only one ECG lead.

The results of the whole system can be seen in Table 7.7. The proposed system has a characteristic of the global classifier trained in an unsupervised manner, which does not require labeled data for training, and consequently, it can be adapted to the individual patient and utilized within the wearable system. The synapse circuit consumes in the worst case an energy of 35 pJ per synaptic event [25]. The soma necessitates significant current to create adequate positive feedback; the level is defined by the maximum synaptic current multiplied with the square root of the number of inputs. The digital spikes generated by the conductance-based neuron circuit are very

Table 7.5 Detection accuracy per beat type

Beat Type	Undetected Beat Percentage
Normal beat	3.44
Left bundle branch block beat	0.19
Right bundle branch block beat	1.42
Atrial premature beat	0.24
Nodal (junctional) premature beat	0
Supraventricular premature beat	0
Premature ventricular contraction	3.87
Fusion of ventricular and normal beat	2.49
Atrial escape beat	0
Nodal (junctional) escape beat	2.62
Ventricular escape beat	1.89
Paced beat	0.08

Table 7.6 Clustering accuracy with number of beats dominant and non-dominant in their clusters

Beat Type	Dominant	Non-dominant	Accuracy
Normal beat	71,146	1072	98.52%
Left bundle branch block beat	7880	180	97.77%
Right bundle branch block beat	6809	347	93.8%
Atrial premature beat	1809	731	71.22%
Nodal (junctional) premature beat	44	39	53.01%
Premature ventricular contraction	6110	738	89.22%
Paced beat	3580	37	98.98%
Ventricular flutter wave	216	36	85.71%

Table 7.7 Clustering accuracy over the whole database

Beat Type	Accuracy in Percentage
Normal beat	95.13
Left bundle branch block beat	97.56
Right bundle branch block beat	93.8
Atrial premature beat	71.05
Premature ventricular contraction	85.77
Paced beat	98.9

narrow (~250 ns). The circuit average power consumption during this period is 2.1 pJ/spike [26]. The average power dissipation measured throughout the whole current integration and action-potential generation phase is 147 pJ over 100 ms. Assuming the worst-case scenario that two beats need to be clustered each second, on average, 200 spikes per second are fired by neurons and 3k synaptic events are required to pass the spikes, leading to the estimated ~100 nW of power. The feature extraction consumes an estimate of 0.7 μW of power [27]. In total, power consumption of the whole classification system including support circuits is less than 1 μW.

7.4 Conclusion

In this chapter, a novel ultra-low-power ECG beat-type classification system for arrhythmia detection has been proposed. In the system, an adaptive ECG interval extraction is utilized for feature extraction and correlation matrix for unsupervised feature selection. The classification of the resulting features is performed with neuromorphic liquid-state machine. The proposed system has a characteristic of the global classifier trained in an unsupervised manner, which does not require labeled data for training, and consequently, it can be adapted to the individual patient. For compatibility with the wearable devices, the classification system requires only one ECG lead. The system as a whole is estimated to consume less than 1 μW, and has an overall clustering accuracy of 95.5%. The performances achieved are comparable with those reported in the literature for fully automated, multi-lead algorithms.

References

[1] G. B. Moody and R. G. Mark, "The mit-bih arrhythmia database on cd-rom and software for use with it," in Computers in Cardiology 1990, Proceedings., pp. 185–188, IEEE, 1990.
[2] J. Rodrıguez-Sotelo, E. Delgado-Trejos, D. Peluffo-Ordonez, D. Cuesta-Frau, and G. Castellanos-Domınguez, "Weighted-pca for unsupervised classification of cardiac arrhythmias," in Engineering in Medicine and Biology Society (EMBC), 2010 Annual International Conference of the IEEE, pp. 1906–1909, IEEE, 2010.
[3] S. Mukhopadhyay and G. Ray, "A new interpretation of nonlinear energy operator and its efficacy in spike detection," IEEE Transactions on Biomedical Engineering, vol. 45, no. 2, pp. 180–187, 1998.
[4] J. Pan and W. J. Tompkins, "A real-time qrs detection algorithm," IEEE Transactions on Biomedical Engineering, no. 3, pp. 230–236, 1985.
[5] W. Zong, G. Moody, and D. Jiang, "A robust open-source algorithm to detect onset and duration of qrs complexes," in Computers in Cardiology, 2003, pp. 737–740, IEEE, 2003.

[6] C. Li, C. Zheng, and C. Tai, "Detection of ecg characteristic points using wavelet transforms," IEEE Transactions on Biomedical Engineering, vol. 42, no. 1, pp. 21–28, 1995.

[7] P. Laguna, R. Jańe, and P. Caminal, "Automatic detection of wave boundaries in multi-lead ecg signals: Validation with the cse database," Computers and Biomedical Research, vol. 27, no. 1, pp. 45–60, 1994.

[8] Y. Sun, K. L. Chan, and S. M. Krishnan, "Characteristic wave detection in ecg signal using morphological transform," BMC Cardiovascular Disorders, vol. 5, no. 1, p. 28, 2005.

[9] I. Guyon and A. Elisseeff, An Introduction to Feature Extraction, pp. 1–25. Berlin, Heidelberg: Springer Berlin Heidelberg, 2006.

[10] C. Wen, M.-F. Yeh, and K.-C. Chang, "Ecg beat classification using greyart network," IET Signal Processing, vol. 1, no. 1, pp. 19–28, 2007.

[11] D. Azariadi, V. Tsoutsouras, S. Xydis, and D. Soudris, "Ecg signal analysis and arrhythmia detection on iot wearable medical devices," in Modern Circuits and Systems Technologies (MOCAST), 2016 5th International Conference on, pp. 1–4, IEEE, 2016.

[12] N. Bayasi, T. Tekeste, H. Saleh, B. Mohammad, A. Khandoker, and M. Ismail, "Low-power ecg-based processor for predicting ventricular arrhythmia," IEEE Transactions on Very Large Scale Integration (VLSI) Systems, vol. 24, no. 5, pp. 1962–1974, 2016.

[13] D. Cuesta-Frau, M. O. Biagetti, R. A. Quinteiro, P. Mico-Tormos, and M. Aboy, "Unsupervised classification of ventricular extrasystoles using bounded clustering algorithms and morphology matching," Medical & Biological Engineering & Computing, vol. 45, no. 3, pp. 229–239, 2007.

[14] P. De Chazal, M. O'Dwyer, and R. B. Reilly, "Automatic classification of heartbeats using ecg morphology and heartbeat interval features," IEEE Transactions on Biomedical Engineering, vol. 51, no. 7, pp. 1196–1206, 2004.

[15] M. Lagerholm, C. Peterson, G. Braccini, L. Edenbrandt, and L. Sornmo, "Clustering ecg complexes using hermite functions and self-organizing maps," IEEE Transactions on Biomedical Engineering, vol. 47, no. 7, pp. 838–848, 2000.

[16] S.-Y. Lee, J.-H. Hong, C.-H. Hsieh, M.-C. Liang, S.-Y. C. Chien, and K.-H. Lin, "Low-power wireless ecg acquisition and classification system for body sensor networks," IEEE Journal of Biomedical and Health Informatics, vol. 19, no. 1, pp. 236–246, 2015.

[17] Q. Long, Y. Ren, J. Han, and X. Zeng, "Vlsi implementation for r-wave detection and heartbeat classification of ecg adaptive sampling signals,"

in Solid-State and Integrated Circuit Technology (ICSICT), 2016 13th IEEE International Conference on, pp. 1597–1599, IEEE, 2016.
[18] J. L. Rodrıguez-Sotelo, D. Cuesta-Frau, and G. Castellanos-Dominguez, "Unsupervised classification of atrial heartbeats using a prematurity index and wave morphology features," Medical & Biological Engineering & Computing, vol. 47, no. 7, pp. 731–741, 2009.
[19] Y. Sun, K. Chan, S. Krishnan, and D. Dutt, "Unsupervised classification of ecg beats using a mlvq neural network," in Engineering in Medicine and Biology Society, 2000. Proceedings of the 22nd Annual International Conference of the IEEE, vol. 2, pp. 1387–1390, IEEE, 2000.
[20] A. Walinjkar and J. Woods, "Personalized wearable systems for real-time ecg classification and healthcare interoperability: Real-time ecg classification and fhir interoperability," in Internet Technologies and Applications (ITA), 2017, pp. 9–14, IEEE, 2017.
[21] G. M. Wojcik and W. A. Kaminski, "Liquid state machine built of hodgkin–huxley neurons and pattern recognition," Neurocomputing, vol. 58, pp. 245–251, 2004.
[22] P. J. Rousseeuw, "Silhouettes: a graphical aid to the interpretation and validation of cluster analysis," Journal of Computational and Applied Mathematics, vol. 20, pp. 53–65, 1987.
[23] N. Otsu, "A threshold selection method from gray-level histograms," IEEE Transactions on Systems, Man, and Cybernetics, vol. 9, no. 1, pp. 62–66, 1979.
[24] D. Arthur and S. Vassilvitskii, "k-means++: The advantages of careful seeding," in Proceedings of the Eighteenth Annual ACM-SIAM Symposium on Discrete Algorithms, pp. 1027–1035, Society for Industrial and Applied Mathematics, 2007.
[25] X. You, "Full-Custom Multi-Compartment Synaptic Circuits in Neuromorphic Structures," Master's thesis, TU Delft, the Netherlands, 2017.
[26] E. Stienstra, "A 32 $\times$ 32 Spiking Neural Network System On Chip," Master's thesis, TU Delft, the Netherlands, 2017.
[27] N. Bayasi, T. Tekeste, H. Saleh, B. Mohammad, and M. Ismail, "A 65-nm low power ecg feature extraction system," in Circuits and Systems (ISCAS), 2015 IEEE International Symposium on, pp. 746–749, IEEE, 2015.

[28] K. O. Stanley and R. Miikkulainen, "Evolving neural networks through augmenting topologies," Evolutionary computation, vol. 10, no. 2, pp. 99–127, 2002.

[29] J.-P. Pfister and W. Gerstner, "Triplets of spikes in a model of spike timing-dependent plasticity," Journal of Neuroscience, vol. 26, no. 38, pp. 9673–9682, 2006.

8

Multi-Compartment Synaptic Circuit in Neuromorphic Structures

Xuefei You

Delft University of Technology, Delft, The Netherlands

Synaptic dynamics is of great importance in realizing biophysically accurate neural behaviors and efficient synaptic learning in neuromorphic integrated circuits. An insight is given into the design space of synaptic dynamics in regards to computation efficiency and biological fidelity. In this chapter, we propose a current-based synapse structure with multi-compartment receptor elements (AMPA, NMDA, and GABAa receptors) and a weight-dependent learning algorithm. The designed circuit offers distinctive dynamic features of receptor elements as well as a joint synaptic function. A cross-correlation methodology is applied to a two-layer recurrent neural network built by multi-compartment receptors to demonstrate the proposed synapse structure. An increased computation ability is verified through temporal synchrony detection among the neural layers in a noisy environment. The design implemented in TSMC 65 nm CMOS technology consumes 1.92, 3.36, 1.11, and 35.22 pJ per spike event of energy for AMPA, NMDA, GABAa receptors, and the advanced learning circuit, respectively.

8.1 Introduction

8.1.1 Synapse

Synapse is an abstract concept proposed to describe the structure that enables signal transmission through neurons. Specifically, it is found at non-continuous joints between adjacent neurons with the presynaptic part located in axon terminals and the postsynaptic part on dendrites (see Figure 8.1).

When action potentials approach the axon terminals, ion channels on the membrane are opened, permitting small lipid bilayer vesicles in the axon,

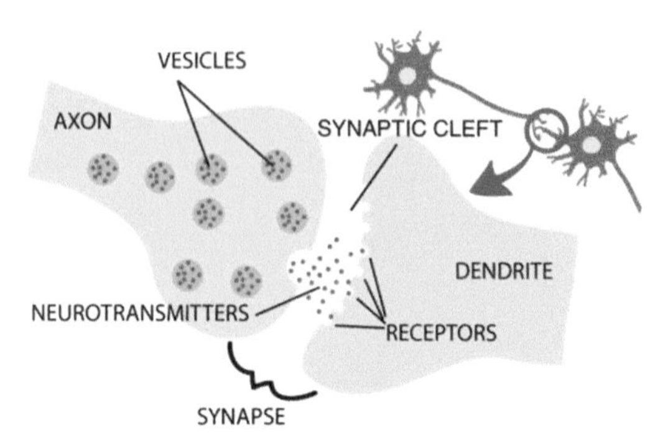

Figure 8.1 Detailed structure of synapse.

called synaptic vesicles, containing an enormous amount of neurotransmitters to efflux into the synaptic cleft. Fusion of vesicles with the membrane allows transmitters inside to be released into the synaptic cleft. Due to the concentration gradient, the transmitters will diffuse toward the dendrites on the postsynaptic neurons and bind to the corresponding receptors via ligands (there are other types of activation mechanisms though), inducing the activation of the particular receptor channels. Only then, the intracellular and extracellular ions are free to flow in between, generating action potentials in the postsynaptic neurons.

8.1.1.1 Synaptic plasticity

The concept of synaptic weight quantifies the learning output. The synaptic weight is potentiated or depressed depending on the analysis, or "learning" of the current cell activities. This kind of weight adaptation ability is called synaptic plasticity. The underlying principle is the Hebbian theory [1]:

Let us assume that the persistence or repetition of a reverberatory activity (or "trace") tends to induce lasting cellular changes that add to its stability. When an axon of cell A is near enough to excite a cell B and repeatedly or persistently takes part in firing it, some growth process or metabolic change takes place in one or both cells such that A's efficiency, as one of the cells firing B, is increased.

Thus, the synaptic weight of the synapses having the same trend as the local network is increased, while that of the irrelevant ones is decreased. This selectivity quantifies as a learning capacity to synapses. Depending on the timescale, the synaptic plasticity is classified into two groups: short-term plasticity (STP) and long-term plasticity (LTP), covering a time range of tens of milliseconds to a few minutes and from minutes to hours, respectively [2].

8.1.1.2 Synaptic receptors

Depending on different ligand types, the effect of transmitter and receptor pairs on the postsynaptic neurons can be either excitatory or inhibitory, corresponding to positive and negative current flow to postsynaptic neurons. Different types of receptors display different temporal dynamics due to their distinctive conducting mechanisms (see Figure 8.2). A generic synapse structure does not capture the diverse temporal dynamics of different types of receptors in biological synapses, which are essential for realization of bio-physically accurate neural behaviors in SNNs (for details, see Section 8.2.2). For this reason, two types of glutamatergic receptors (AMPA and NMDA) and one GABAergic receptor (GABAa) are discussed in detail below as the foundation of further hardware implementations.

8.1.1.2.1 *AMPA receptor*

The α-amino-3-hydroxy-5-methyl-4-isoxazolepropionic acid receptor, known as AMPA receptor, is one of the most common receptors in the nervous system. Mostly, the AMPA receptor is permeable to sodium (Na^{+}) via ion channels. Upon binding of transmitters on AMPA receptors, positively charged Na^{+} enters the AMPA ion channels and depolarize the cell, thus inducing action potentials. However, the maximum conductance of AMPA receptors is limited by the intracellular calcium (Ca^{2+}) concentration. The prevention of calcium entry into the cell is reported to protect against excito-toxicity [4]. AMPA receptors open and close quickly due to a straightforward

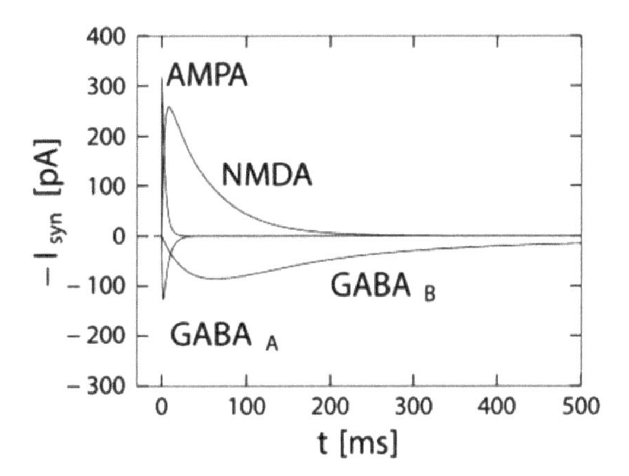

Figure 8.2 Temporal dynamics of four types of receptors [3].

mechanism of channel opening and closing, and are thus responsible for fast signal transmission [5].

8.1.1.2.2 *NMDA receptor*

The ion channels of the N-methyl-D-aspartate receptor, also named NMDA receptor, is voltage-dependent, which is distinctive compared with other glutamatergic receptors. This dependency initially arises from the non-selectivity of its ion channels. When ligand-binding occurs, the non-selective ion channels are open to extracellular magnesium (Mg^{2+}) and zinc (Zn^{2+}), which will bind to specific sites on the receptor and block the channels for any other ions. To eliminate this blockage, a certain level of depolarization of the cell is necessary, usually through the influx of Ca^{2+} [6]. Once cleared, the ion channels introduce both Ca^{2+} and Na^{+} into the target cell. At the same time, in response to the increased level of depolarization, more AMPA receptors are inserted into the membrane, creating more possibility of ion influx. Thus, the conductance of NMDA receptor has a boost effect on the postsynaptic current.

To activate NMDA receptors, the presynaptic activities should introduce free transmitter to the dendrites while the postsynaptic depolarization should, as a prerequisite, open the ion channels on the receptors. This kind of dual function of pre- and postsynapses implies the role of NMDA receptor in synchrony detection and biological emulation. On the temporal aspect, the NMDA receptors are typically three to six times slower than AMPA with regard to synaptic dynamics [7], due to a more complex binding mechanism and small unchanneling speed.

8.1.1.2.3 *GABA receptor*

The gamma-Aminobutyric acid receptor, also called GABA receptor, is a primary inhibitory channel carrier in the nervous system. Two classes of the GABA receptors are defined according to different activation mechanisms, ligand-gated GABAa receptor, and protein-coupled GABAb receptor. A significant stimuli intensity is required to evoke the GABAb-mediated responses, which is hard to achieve in biological experiments or further obtain the estimation of GABAb receptors [7].

The GABAa receptor is permeable to chloride (Cl^{-}). When activated, the GABAa receptor conducts Cl^{-} through the ion channels, causing the hyperpolarization of the cell and a lower possibility of neural firing. This inhibition function of the GABAa receptor is a prerequisite for balancing excitation and inhibition, thus stabilizing neural network [8]. The GABAa

Table 8.1 Biological dynamics for three receptors [22, 26]

	Polarity	Rise and Fall Times (ms)	Conduction Remarks
AMPA	+	0.4–0.8, 5	1-step, fast EPSC
NMDA	+	20, 100	2-step, voltage dependency, slow EPSC
GABAa	–	3.9, 20	1-step, fast IPSC

receptors have similar temporal dynamics to AMPA, i.e., both the rise and the fall times of EPSCs are comparable. A comparison of temporal dynamics of the above receptors is shown in Table 8.1.

8.2 Model Extraction

8.2.1 Model of the Synapse

One of the basic and direct models describing the synaptic conductance properties is the exponential decay model where the rising phase of the synaptic conductance is assumed to be instant [9], i.e., the release of transmitters, its corresponding diffusion across the cleft, the receptor binding, and channel opening all occur very fast. The conductance of the synapse at time t is then:

$$g_{syn}(t) = \bar{g}_{syn} \cdot e^{-(t-t_0)/\tau} \cdot \Theta(t - t_0) \tag{8.1}$$

where $\bar{g}_{syn}$ is the maximum conductance of the synapse, t_0 is the onset time of the presynaptic spike while τ is the decaying time constant, and $\Theta(x)$ is the Heaviside step function.

The synaptic weight is represented by $\bar{g}_{syn}$ after certain learning, while $e-(t-t_0)/\tau$ induces biologically analogous exponential conversion. For some IPSCs, the exponential decay model is validated to outline their activities, because the rising phase of these currents is much shorter compared with the decaying phase, like the GABAa-induced currents (see Table 8.1). However, when it comes to EPSC induced by NMDA receptors, which has comparable temporal dynamics in both rising and decaying phases, the model fails to emulate its behaviors. A more detailed model with two separate exponential components is introduced:

$$g_{syn}(t) = \bar{g}_{syn} \cdot f \cdot (e^{-(t-t_0)/\tau_{decay}} - e^{-(t-t_0)/\tau_{rise}}) \cdot \Theta(t - t_0) \tag{8.2}$$

The factor *f* is used to normalize the total amplitude of the sum to $\bar{g}_{syn}$, and τ decay and τ rise are the time constants for decaying and rising phases, respectively. The neuromorphic design for this model is more complex due to one extra rising phase consideration.

The transistor operating in the sub-threshold region is of the most interest mainly for two reasons: (*i*) first, due to its exponential relationship between the drain currents and the gate voltages analogous to the biological model, and (*ii*) second, due to its extremely low power consumption. When $V_{ds} \geq 4U_T \approx 100$ mV, the transistor enters the sub-threshold saturation region where a pure exponential relationship between V_{gs} and I_{ds} is generated:

$$I_{ds} = I_0 \cdot e^{\kappa_n V_{gs}/U_T} \tag{8.3}$$

where I_0 is the current value when V_{gs} equals the threshold voltage of the transistor, acting as a current scale control, and κ_n represents the sub-threshold slope for n-type MOSFET. V_{gs} and U_T are the gate-to-bulk and threshold voltages of the transistor.

Regardless of the distinctions among different synapses, a general structure can be decomposed into two functional units: the synaptic learning block inducing the weight adaptation and the distinctive receptor compartments bringing in chemical ligand-gated channel control. In next subsection, the mathematic models of the synaptic learning rules will be explained.

8.2.2 Learning Rules

Synaptic plasticity models of various complexity levels [10] have been proposed in regards to various application requirements, ranging from the abstract ones to the more biologically realistic and detailed ones. In this subsection, two popular algorithms are explained: pair-based spike-timing-dependent plasticity (PSTDP) and triplet-based STDP (TSTDP).

8.2.2.1 Pair-based STDP

PSTDP, an almost symmetrical pattern of Hebbian's theory, is a learning process that can adapt the synaptic weight according to temporal correlations between pre- and postspikes of a target synapse. These correlations should be within milliseconds time range in accordance with biological temporal features: if the pre spike precedes the post spike, a potentiation of the synaptic weight occurs; in contrast, if a reversed sequence happens, depression is induced. A diagram of the STDP learning function is shown in Figure 8.3. Two factors of concern in this learning window are time constants (τ) and amplitudes (A) of two phases. The time constant indicates the temporal range

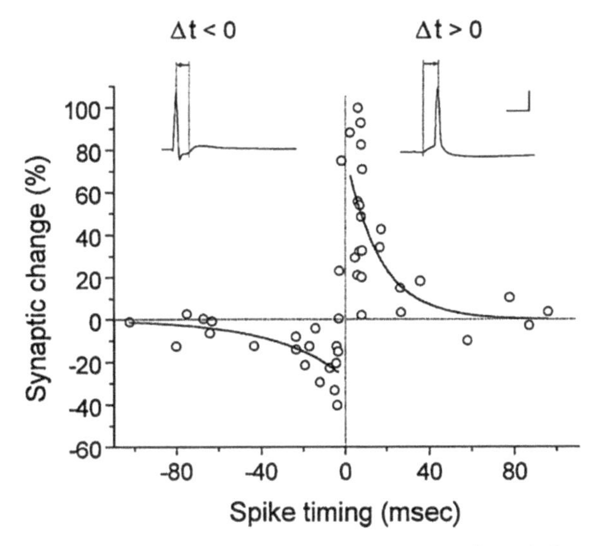

Figure 8.3 The learning window of STDP learning rule. The hollow circles are experimental data of EPSC amplitude percentage change at 20–30 min after repetitive stimuli of pre- and postspikes at a frequency of 1 Hz [25]. The spike timing is defined as the temporal interval between post- and prespikes. An exponential fit of those data points is outlined with two smooth curves (LTP and LTD). For LTP and LTD, respectively, A = 0.777 and 0.273; τ = 16.8 and 33.7 ms.

where the correlation happens while the amplitude controls the adaptation level. The STDP rule is expressed as below:

$$\Delta w^{+} = A^{+} \cdot e^{-\Delta t/\tau_{+}} \;\; \Delta t > 0 \tag{8.4}$$

$$\Delta w^{+} = A^{+} \cdot e^{-\Delta t/\tau_{+}} \;\; \Delta t < 0 \tag{8.5}$$

where Δt is the temporal difference between a single pair of post- and pre spikes. A_{+} and A_{1} are the maximum amplitudes, while τ_{+} and τ_{-} are time constants of the potentiation and the depression phase, respectively. These parameters impact the area of the weight update curves during potentiation and depression. It is observed that stable learning is realized when the aggregate area of depression exceeds that of potentiation in the weight update function.

In contrast, weaker depression results in the extreme potentiation of synaptic weights and the eventual shorting of outputs to inputs. The biological experiments [11] illustrate that LTP is strengthened with relatively small initial weight, while LTD does not exhibit too large dependency. For this

Table 8.2 Weight dependency rules

	Additive	Multiplicative	Power law
+	$A = \lambda$	$A(w) = \lambda(1 - \omega)$	$A(w) = \lambda(1 - \omega)^{\mu}$
−	$A = \lambda\alpha$	$A(w) = \lambda\alpha\omega$	$A(w) = \lambda\alpha\omega^{\mu}$

reason, specific initial weight dependence rules are incorporated in synapse models, e.g., the additive [12], the multiplicative [13], and power law update rule [14]. The detailed weight dependency rules are listed in Table 8.2. The synaptic weight is denoted by w ($0 < w < 1$), $\lambda \leq 1$ is the learning rate, α is an asymmetrical parameter, and μ drives the combination point between additive and multiplicative models. For example, the learning function in (8.4) and (8.5) does not involve weight dependence in both A_+ and A_-, which gives it additive update features [10].

8.2.2.1.1 *Triplet-based STDP*

Derived from the pair-based STDP, triplet-based STDP (TSTDP) incorporates the correlations among three consecutive spikes. The mathematical representation of TSTDP learning rule is given by:

$$\begin{aligned} \Delta w^{+} &= e^{-\Delta t_1/\tau_+}(A_2^{+} + A_3^{+} \cdot e^{-\Delta t_2/\tau_y}) \qquad \Delta t > 0 \\ \Delta w^{-} &= e^{\Delta t_1/\tau_-}(A_2^{-} + A_3^{-} \cdot e^{-\Delta t_3/\tau_x}) \qquad \Delta t < 0 \end{aligned} \tag{8.6}$$

where τ_+, τ_-, τ_y, and τ_x are time constants concerning triplet spikes (details shown in Figure 8.4). A_2^+ and A_2^- are second-order potentiation and depression amplitude parameters, while A_3^+ and A_3^- are third-order amplitudes. Δt_1, Δt_2, and Δt_3 represent $t_{post}(n) - t_{pre}(n)$, $t_{post}(n) - t_{post}(n - 1) - \varepsilon$ and $t_{pre}(n) - t_{pre}(n - 1) - \varepsilon$, respectively, with ε a small positive constant to ensure that the weight update uses the correct values occurring just before the target pre- or postsynaptic spike.

While the PSTDP can capture the basic learning algorithms of the synapse, it might be insufficient to demonstrate the frequency effect and higher-order outcomes observed in biological models [15]. The PSTDP is

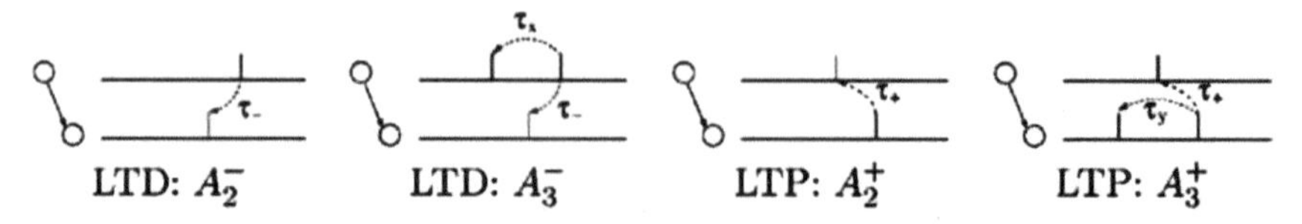

Figure 8.4 Time constant parameter interpretation of TSTDP.

a substantially linear model where the overall effect of consecutive spikes may sometimes counteract with each other, thus restricting the frequency effect of the synapse. In contrast, this nonlinearity is demonstrated in the third-order spike patterns. The TSTDP is capable of reproducing triplet and quadruplet outcomes in experiments; unfortunately, the design complexity increases with the higher biological similarity, leading to the redundancy in hardware implementation.

8.3 Component Implementations

This section introduces three main types of synaptic learning circuits applying the learning rules discussed in Section 8.2.2. The circuits are implemented in UMC 65 nm technology. Due to the fact that the biological signals usually span over milliseconds to seconds range, one of the detrimental elements interfering the synaptic performance is the spontaneous leakage of transistors and capacitors. Consequently, we utilized a low leakage transistor (LL) technology.

8.3.1 Learning Rule 1: Classic STDP

As shown in Figure 8.5(a), the circuit [16] has a main branch (M1–M6) responsible for pulling up or down the synaptic weight stored in a weight capacitor C_w. Additionally, two leaky integrators are implemented to offer the controllability of both potentiation and depression windows (V_{pot} and V_{dep}) in typical learning window (see Figure 8.3).

A complete circuit design is illustrated in Figure 8.5(b). The potentiation current Ipot is initially drained from transistor M1, which is biased by V_{pot}. V_{pot} is rested at high rail voltage for most of the time but is pulled down via charging from M9 when a prespike pulse approaches for a short period of time. After that, V_{pot} slowly discharges through M7-M8. The discharge occurs in an almost linear pattern if the channel length modulation effect is ignored. V_{tp} controls the discharge speed, thus determining the decay time constant of potentiation phase τ_+ in the learning window in Section 8.2.2. To induce exponential dynamics, transistor M1 is maintained in the saturated sub-threshold region characterized by stable exponential dynamics ($V_{ds} >$ 100 mV). At certain point of the decay process, a post spike is transmitted and further activates the upper main branch. The generated current from upper branch is injected to C_w, causing an increment in synaptic weight V_w. V_p acts as a controllable source, adjusting the amount of current to be allowed to flow

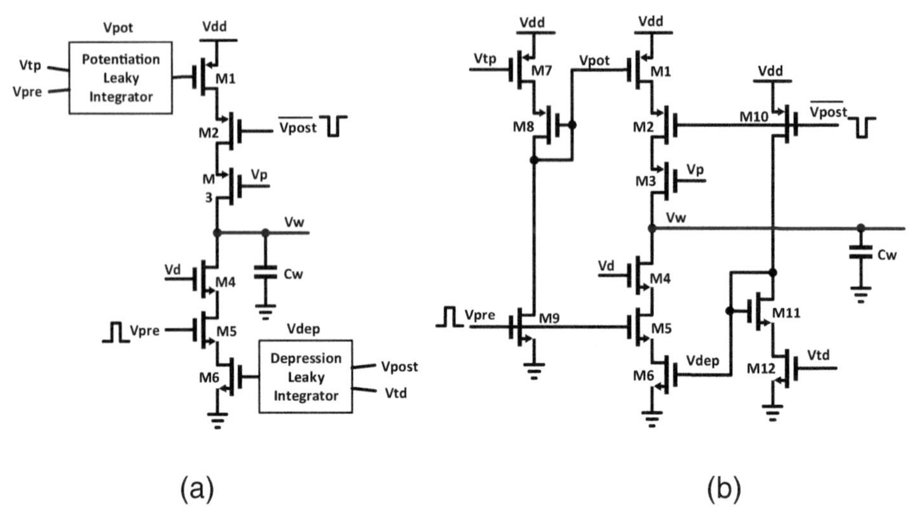

Figure 8.5 The simplified structure diagram of the classic PSTDP circuit [16] (a) with the leaky integrators represented by squares boxes. (b) The detailed circuit.

into C_W, thus determining the amplitude (A_+) of the potentiation window. The same mechanism applies to the depression phase of the synapse with a complementary design.

This PSTDP circuit offers a low power consumption (activated only when spike occurs) and simplicity. Although regarded as a typical PSTDP model, the design faces one detrimental problem: the leaky integrators in both potentiation and depression phases cannot guarantee a voltage range that drives M1 and M6 into the sub-threshold region, i.e., this circuit cannot offer an exponential dynamics expected in real neural cells, thus reducing its biological fidelity.

8.3.2 Learning Rule 2: Advanced STDP

The circuit [17] in Figure 8.6 partially solves the problem mentioned above. The general idea of pulling up or down the synaptic weight through injection or efflux of activated currents from the weight capacitor is similar to that shown in Figure 8.6(a). The main difference is that the amplitude control block is incorporated into the so-called advanced leaky integrator circuits (blue dashed blocks in Figure 8.6(b)) representing both potentiation and depression learning windows. An additional weight dependence circuit (red dashed line block in Figure 8.6(b)) complements biological features.

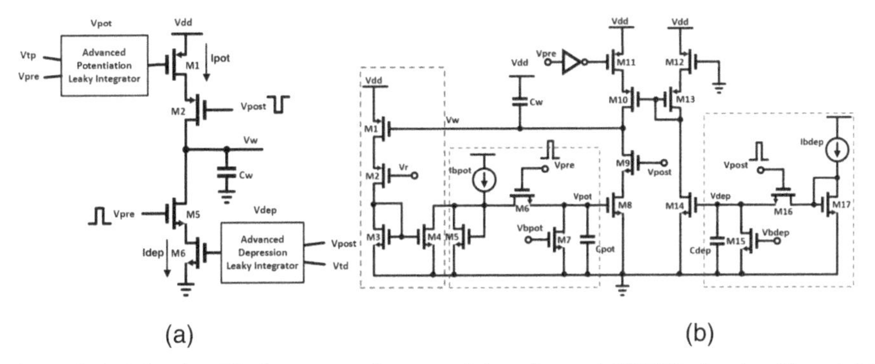

Figure 8.6 The simplified structure diagram of the advanced PSTDP circuit with weight dependence [17]. (a) with the advanced leaky integrators represented by square boxes. (b) Circuit details.

Unlike the circuit in Figure 8.5, the bias voltage V_{pot} loaded to the main branch upon the arrival of prespike is adjustable through a controllable current source I_{bpot}. Again, the postspike within the time range of the learning window activates I_{pot} and removes charges stored in C_w at certain point of time. It should be noted that this circuit follows a complementary design flow, i.e., the current efflux from C_W means an increment of synaptic weight. The time constant of the learning window is determined by M7 biased by V_{bpot}. Without the interference of the amplitude control on main branch, this circuit is able to achieve desirable exponential relation of synapses. The same analysis applies to the depression phase on the right side. A notable simplification is that the advanced leaky integrator activated by postspikes is shared by all the synapses connected to the target neuron. The depression currents for individual dendritic synapses are generated via duplicated current branches, allowing large area saving in VLSI network.

The circuit includes the weight dependence feature (built by M1–M4) in addition to the basic STDP learning rule. M1 is a low-gain transistor operating in the strong inversion region, while M2 is a high-gain transistor operating in the sub-threshold region. To enlarge the possible synaptic weight range, a low-threshold-voltage transistor is used or M1. The bias voltage V_r tunes the weight dependence level. When V_W decreases, i.e., the synaptic weight increases, the current through M1–M2 increases. M1 will be driven to work in the linear region by M1, which means that a change of synaptic weight is converted linearly to a current subtraction from I_{bpot}. M5, a diode connected transistor, works in the sub-threshold as M5 and M8 share the same bias voltage upon the arrival of a presynaptic spike, and M8 has to be maintained in the sub-threshold region to offer an exponential relation. The linear decrement of current in M5 is thus mapped to M8 through a current mirror mechanism. In this way, a multiplicative weight dependence (see Table 8.2) is achieved by adding four extra transistors.

An extra continuously activated current branch (M1–M3), especially in a design where most of the transistors conduct only upon the presence of pre- or postspikes, will cause a much larger power consumption in comparison with the circuit without the weight dependence.

8.3.3 Learning Rule 3: Triplet-Based STDP

From (8.4)–(8.7), it can be observed that the second-order components in the TSTDP model are the same as in the PSTDP model. The distinction of TSTDP arises from the third-order components which is summed together with the amplitude components (A_2^+ and A_2^-) before an exponential conversion. Based on this observation, the second-order design in Section 8.2.2 can be well used in TSTDP implementations, whereas extra third-order implementations need to be supplemented. In the classic PSTDP circuit, the amplitude control of both potentiation and depression learning window is blended into the main branch, namely the amplitudes cannot be separated to be further added to the third-order components. The advanced PSTDP with separate tuning blocks of learning window is a feasible method.

A triplet-based STDP circuit [18] is shown in Figure 8.7. Four individual parts can be easily identified in the circuit to match with two second-order and two third-order components in triplet learning algorithm. The middle two parts have similar design to the advanced PSTDP circuit (for details, see Section 8.2.2) generating second-order potentiation and depression window

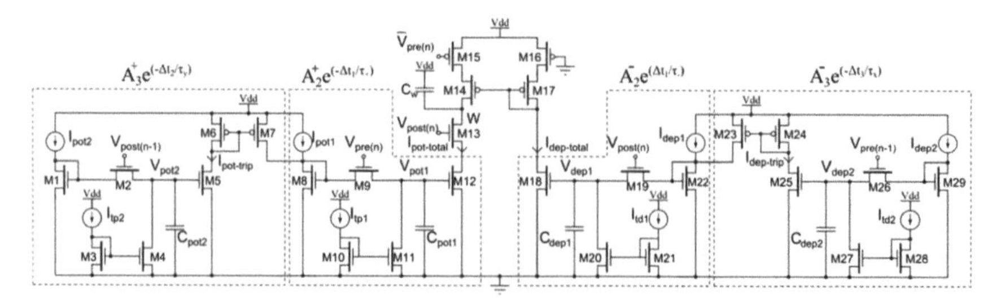

Figure 8.7 Triplet-based STDP circuit [18]. The red and blue dashed blocks represent potentiation and depression learning blocks, respectively.

activated by the current prespike $V_{pre}(n)$ and postspike $V_{post}(n)$, respectively. The leftmost and rightmost parts are built by the same advanced leaky integrator but are activated by the previous prespike $V_{pre}(n - 1)$ and postspike $V_{post}(n - 1)$.

The resultant current from those two blocks are mirrored and summed with the corresponding amplitude control current I_{bpot} and I_{bdep}, introducing the third-order effects into the second-order ones. The incorporation of the triplet spikes enlarges the biological fidelity of the synapse at an expense of a doubled area and power consumption, which should be taken into account in a large-scale network design.

8.3.4 Synaptic Receptors

The synaptic weight update is converted to the synaptic currents, which will be integrated by the postsynaptic neuron. This is achieved through the receptors, i.e., functional integrators transferring the weight information to EPSCs or IPSCs upon upcoming stimuli.

8.3.4.1 AMPA receptor

A differential-pair integrator (DPI) [19] structure is applied to emulate the fast rising and decaying dynamics of AMPA receptor shown in Figure 8.8(a). The receptor potential state V_{syn} stored in C_{syn} is decremented fast with every efflux of charges from M2, M3, and M4 path upon the arrival of presynaptic spikes and then decays linearly toward the high rail from M5. The EPSC is obtained via an output transistor M6 operating in the sub-threshold region. Four characteristics of this circuit should be highlighted:

- The differential structure of M1 and M4 offers a "seesaw-like" control over the currents through these two branches. When the synaptic weight is maintained, the current efflux in paths M2, M3, and M4 can be adjusted through bias M1. The driving ability of the synapses to the postsynaptic neurons is thus flexible depending on diverse configurations between the target neuron and the corresponding dendritic synapses.
- The decaying time constant τ_{AMPA} is tunable through the bias V_{τ} in dealing with the various experimental data obtained in different experiments or receptor locations in the brain.
- The circuit follows a linear differential equation dynamic, enabling direct summation of identical receptor sources.

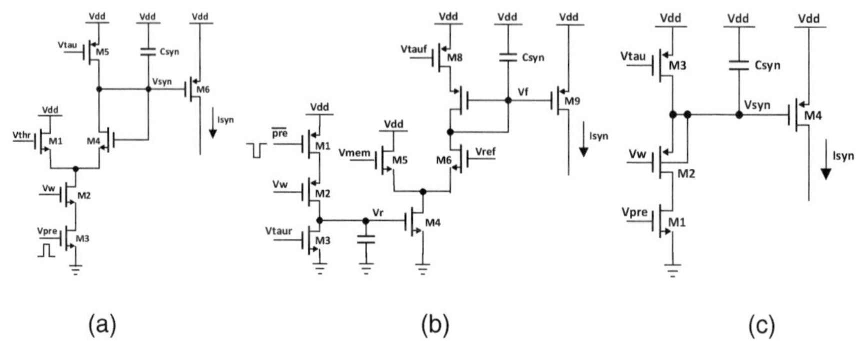

Figure 8.8 Receptor implementations according to their distinctive dynamics [19]. (a) AMPA receptor using DPI structure; (b) NMDA receptor using two-stage conduction mechanism; (c) GABAa using log-domain integrator structure.

- The circuit has a compact structure and low power consumption because the main circuit mostly conducts only in the presence of the presynaptic spikes, which lasts no more than 2 ms.

8.3.4.2 NMDA receptor

Unlike the single exponential dynamics used for AMPA receptor, the charging phase of NMDA receptor cannot be ignored due to its relatively large portion in the whole temporal range. Thus, a double exponential function should be displayed in NMDA receptor design, as well as its distinctive weight dependence. In Figure 8.8(b), the presynaptic spike enables an instantaneous current influx into C_{rise} in the rising phase, the amplitude of which is controlled by V_w. The bias V_{taur} determines the discharge speed of Crise. During this controllable period of time, the transistor M4 is always active, inducing the voltage drop of V_f. After that, capacitor C_{syn} begins to discharge through M8 biased by V_{tauf}, adjusting the falling time constant. In this way, controllable double exponential dynamics are generated.

To incorporate the distinctive voltage dependence of NMDA receptors, a differential pair is added to the circuit, forming a comparison between V_{mem} and V_{mth}. When the postsynaptic neuron is depolarized, V_{mem} surpasses V_{mth}, introducing valid current flux into C_{syn}. On the contrary, if Vmth surpasses V_{mem}, no or a small fraction of current is induced to generate EPSCs.

8.3.4.3 GABA receptors

Due to the similarity between the dynamics of AMPA and GABAa receptors (except for the polarity), a complementary design of AMPA synapse (DPI synapse) can be efficiently applied to induce IPSCs. Three control voltages are needed; however, since the inhibitory synapses do not exhibit learning properties, the inhibitory level is independent on the synaptic weight for which the synaptic weight value is maintained. A log-domain integrator is chosen as the implementation of the GABAa receptor for its simplicity as well as linear dynamics [19] (see Figure 8.8(c)).

8.4 Component Characterizations

In this section, characterizations of the architectures discussed in the previous section are described. First, the learning algorithm should be able to process signals in biologically plausible time range, i.e., of the order of tens of milliseconds. Second, the learning should display an exponential dynamics observed in real neurons. Finally, both power and area consumptions should be maintained as low as possible. Additionally, the distinctive properties of three types of receptors, AMPA, NMDA and GABAa, are demonstrated individually as well as the joint function of combinations in a unit network called a cluster of neural network.

8.4.1 Learning Rule 1: Classic STDP

As described in Section 8.2.1, the circuit adjusts the amplitudes and time constants of the STDP learning window in both potentiation and depression phases through V_{tp}, V_{td}, V_p, and V_d biasing voltages, which influences the learning levels and ranges of correlation between spikes. It gives flexibility of various configurations depending on experimental needs or the development of brain explorations. In Figure 8.9, as an example, several biasing voltages are given to V_{tp} and V_{td}, generating different time constants of the leaky integrator voltages V_{pot} and V_{dep}. The charging and discharging processes in the leaky integrator circuits are stored in the parasitic capacitors set on fF level, which offers an upper bound to the time constants that can be obtained through the leaky integrator blocks. Another detrimental problem is that transistors M1 and M6 operate in the saturated sub-threshold region to offer the plausible exponential dynamics observed in real neurons. The leaky integrator circuit is not able to maintain V_{pot} and V_{dep} in the preferred

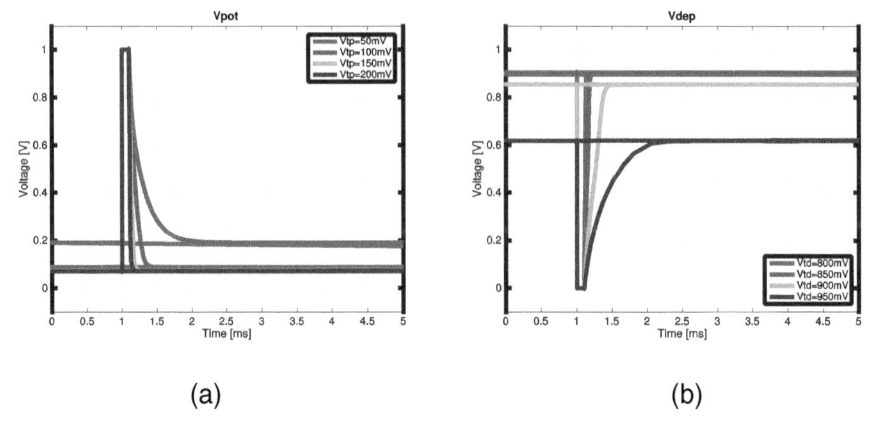

Figure 8.9 Time constant control of Vpot and Vdep in the classic STDP circuit via biasing voltages V_{tp} and V_{td}.

region, i.e., below the threshold voltage of M1 and M6. Thus, the exponential dynamics are absent in the learning algorithm.

8.4.2 Learning Rule 2: Advanced STDP

Unlike the leaky integrator used in the classic STDP circuit, the one in the advanced STDP circuit has an extra capacitor, increasing the load diversity of charging and discharging phases. The time constants are well controlled via the leaky branch M7 and M15 biased by Vbpot and V_{bdep} (see Figure 8.10(a)). The range reaches several tens of milliseconds. The amplitudes of the learning window, on the other hand, are controlled by active current sources I_{bpop} and I_{bdep} (Figure 8.10(b)). For pure temporal coding, an important factor in the stability of any STDP approach is the configuration of maximum facilitation amplitudes A^+ and A^-. These parameters impact the area of the weight update curves during potentiation and depression. It is observed that stable learning is realized when the aggregate area of depression exceeds that of potentiation in the weight update function. On the other hand, weaker depression results in the extreme potentiation of synaptic weights and the eventual shorting of outputs to inputs. This behavior prevents the realization of any practical network transfer function. Compared with the last learning architecture, the advanced STDP circuit has a better controllability over the learning window parameters, and thus it is possible to offer the plausible exponential dynamics, as well as a more stable performance.

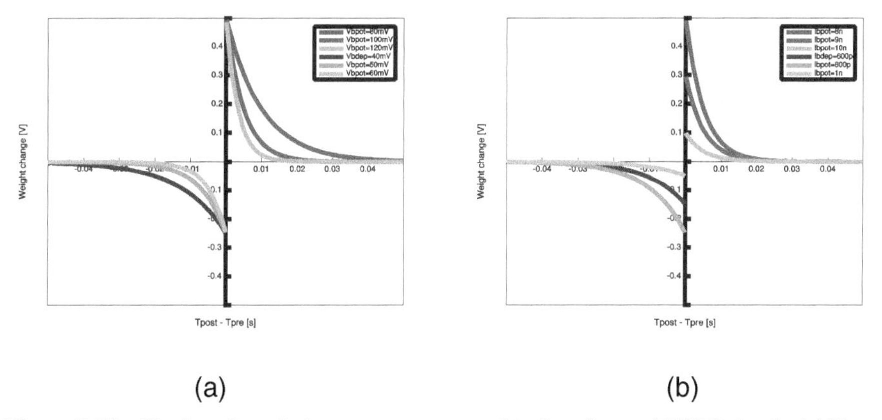

Figure 8.10 The learning window parameter control in the advanced STDP circuit. (a) Time constants adjusted by V_{bpop} and V_{bdep}; (b) amplitude adjusted by I_{bpop} and I_{bdep}.

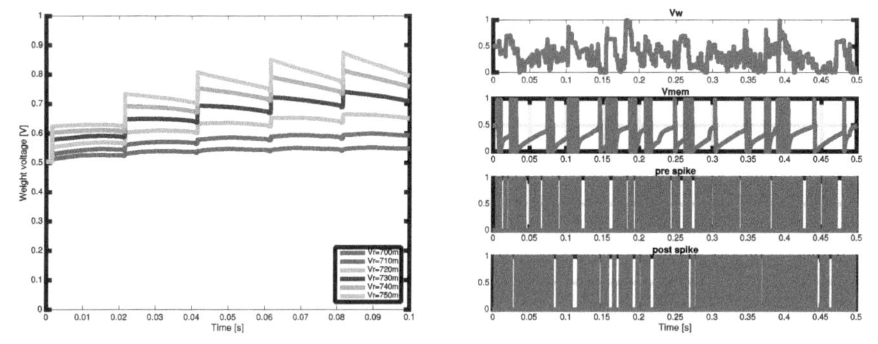

Figure 8.11 (a) Weight dependence property of the advanced STDP circuit. In this example, the presynaptic and postsynaptic signals are of 50 Hz frequency with 1 ms delay. Vr determines the weight dependence level. (b) Sample synaptic weight evolvement and the membrane voltage distribution in the advanced STDP circuit.

Figure 8.11(a) demonstrates the functionality of the extra weight dependence block in the advanced STDP circuit. The input signal pairs induce a stable increment of the synaptic weight value. As the weight adjustment level decreases (V_r increases), this increment becomes less effective, which indicates a lack of adaptability of synaptic weight. This illustrates the bimodal weight distribution in a long period of time reported in [10] for additive weight update rules where the weight dependence is absent. In contrast, the synapse with certain weight dependence, i.e., the multiplicative update rules, shows a unimodal distribution of synaptic weight [10]. An example weight evolvement and the corresponding membrane voltage distribution with Poisson distributed pre- and postsynaptic input signals of 200 Hz are displayed in Figure 8.11(b).

8.4.3 Learning Rule 3: Triplet-based STDP

TSTDP synapses, while suited for pair-based temporal coding, can additionally support triplet and quadratic dynamics, which offers additional nonlinear dynamics to the system, reported to be biologically realistic [11].

Figure 8.12(a) illustrates the weight change induced in the synapse as a function of temporal difference between two postspikes in a triplet. The influence of spike pairs is negated in this analysis using a fixed 2 ms temporal difference in all experimental runs. As observed, the closer the two postspikes, the larger is the effected potentiation in the synapse. Such third-order spike interactions can be observed for temporal differences under 30 ms. Furthermore, the impact of third-order spike interactions wanes, leaving the base

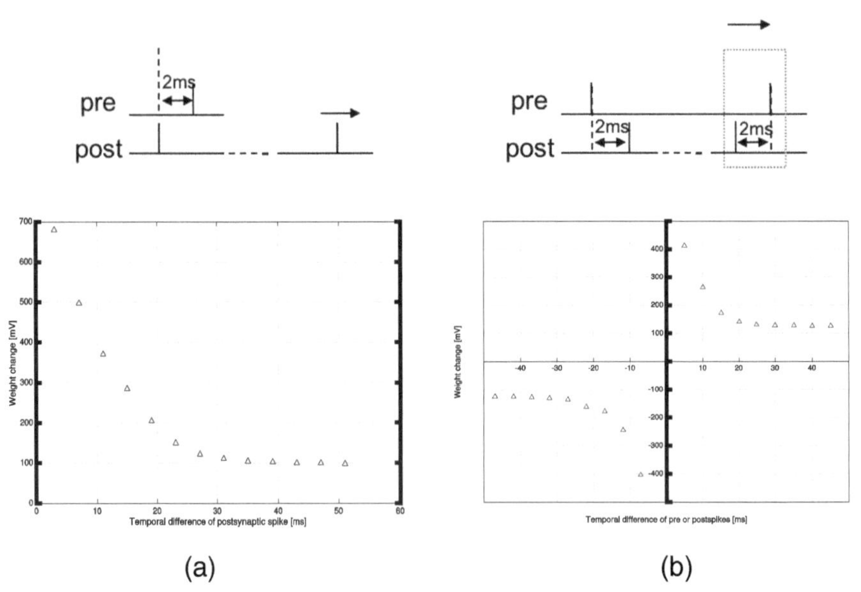

Figure 8.12 Synaptic weight change due to temporal difference between pre- and postspikes in a (a) triplet and (b) quadrant. The insets interpret the temporal settings of input spikes. In (a), the x-axis represents the temporal difference between two postspikes while that in (b) are between post and prespike depending on its in depression or potentiation phases. The dashed blocks in the insets in (b) indicate the temporal shifting block.

Table 8.3 Comparison of three learning rules

	Classic STDP	Advanced STDP	Triplet STDP
Exponential dynamics	×	✓	✓
2nd-order dynamics	✓	✓	✓
Higher-order dynamics	×	×	✓
Weight dependence	×	✓	✓
Time constant range (ms)	0 to several	0–100	0–100
Energy per spike (pJ)	21.64	35.22	82.07
Normalized area	1	3	5

potentiation caused by the 2 ms spike pairs, shown as the flat portion of the curve. Similar experiments are applied to a quadrant except that the shifting component is now another pair of pre- and postspikes (see Figure 8.12(b)). The forth-order effect lasts 20–30 ms for both phases. After this, constant weight changes are observed as the flat portion, caused by the fixed temporal difference pre–post pairs.

A parallel comparison of the above three learning rule circuits is reported in Table 8.3. Note that the energy features are obtained under 200 Hz input spike trains. The areas are an approximate normalized values (the area of classic STDP is set as 1) for preliminary comparison. The classic STDP circuit does not exhibit basic exponential dynamics although the area and energy features are the best among three. The triplet STDP has higher-order dynamics, achieving a better fidelity. The corresponding costs are almost double power and area features as the advanced STDP circuit. In the following experiments, the advanced STDP is chosen for integration due to its best trade-off between various properties. For other certain applications, other models are still possible.

8.4.4 Synaptic Receptors

The synaptic time constant is regulated by transistor M5 biased via V_{tau} to cover various temporal range for AMPA receptors. The possible time constant dynamics are displayed in Figure 8.13(a), ranging from several to tens of milliseconds. Similarly, the time constants for NMDA receptor in both rising and falling phases are adjustable via two separate bias voltages V_{taur} and V_{tauf}, displayed in Figure 8.13(b)(c). Double exponential dynamics are generated, which gives a better fidelity with biological neuron cells.

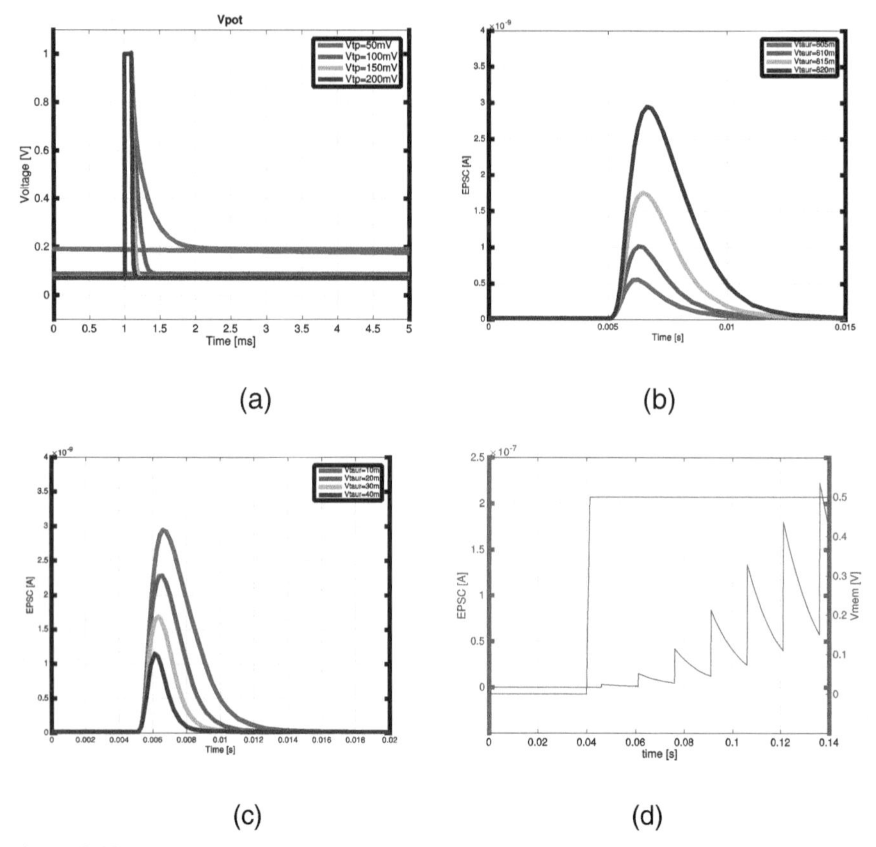

Figure 8.13 Single-receptor characterization. (a) τ_{AMPA} control by V_{tau}; (b) rising-phase time constant control by V_{taur} of NMDA receptor; (c) falling-phase time constant control by V_{tauf} of NMDA receptor; (d) voltage dependence demonstration of NMDA receptor.

The weight dependence of NMDA receptor is demonstrated through a comparison of V_{mem} and a reference voltage V_{ref}. A sequence of presynaptic spikes is introduced to synapse. V_{mem} is a step signal from 0 to 500 mV (larger than V_{ref}) onset time at 40 ms. It can be observed in Figure 8.13(d) that a growing output current starts to emerge from the onset of V_{mem}. A linear increase of EPSC amplitudes can be found at each stimuli, as commonly found in AMPA and GABAa receptors. When stimuli are densely distributed, single NMDA EPSC fails to return to resting line before the next stimuli comes due to large decaying time constants, resulting in a summation behavior of previous activities.

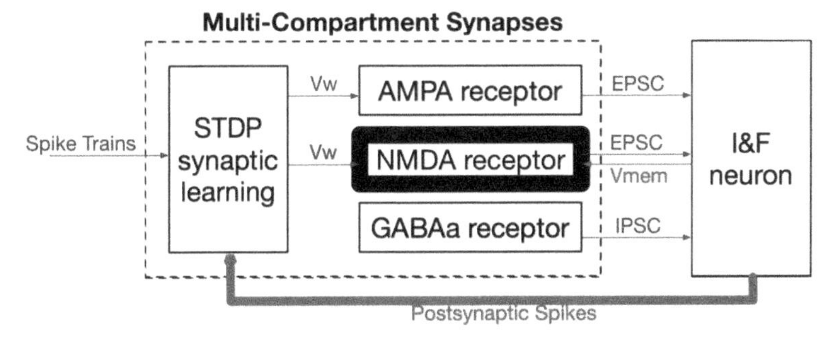

Figure 8.14 Top-level architecture of a cluster network. Multi-compartment synapses consist of three distinctive receptors and a learning circuit inducing adaptable learning results to receptors. The blue line represents the forward transmission signals, while the red ones are feedbacks.

8.4.4.1 Environment settings

The top-level architecture of this network is shown in Figure 8.14. The STDP learning circuit incorporates the current spike activities of the presynaptic and the postsynaptic neuron activities to induce the synaptic weight adaptation. This resultant weight is then transmitted to receptor array (except for the GABAa receptor) to achieve distinctive dynamics. As a weight-dependent receptor, the NMDA receptor is updated according to the membrane state, which is a feedback signal from the postsynaptic neuron. This joint function of three receptors is achieved through the summation of different ions. In the biophysical neuron, the ions flowing through different receptor channels are gathered together in soma body. The overlapping function of those ions are tested at the hillock, which determines whether an action potential is produced, i.e., a summation form of individual responses of different receptors is determined.

The cluster circuit is illustrated in Figure 8.15. The advanced learning circuit with biologically realistic weight dependence as well as a low area and energy consumption feature is introduced in Section 8.2.2, and the receptor architecture is discussed in Section 8.2.4. For the neuron architecture, the classic integrated-and-fire (I&F) neuron model proposed in [20] is applied. The membrane state V_m is stored in C_m, which is charged by the EPSCs or IPSCs generated by the dendritic synapses. The voltage gain is obtained with two cascade inverters with a threshold voltage V_{mth}. This voltage gain block and two capacitors, C_{fb} and C_m, form a positive feedback loop.

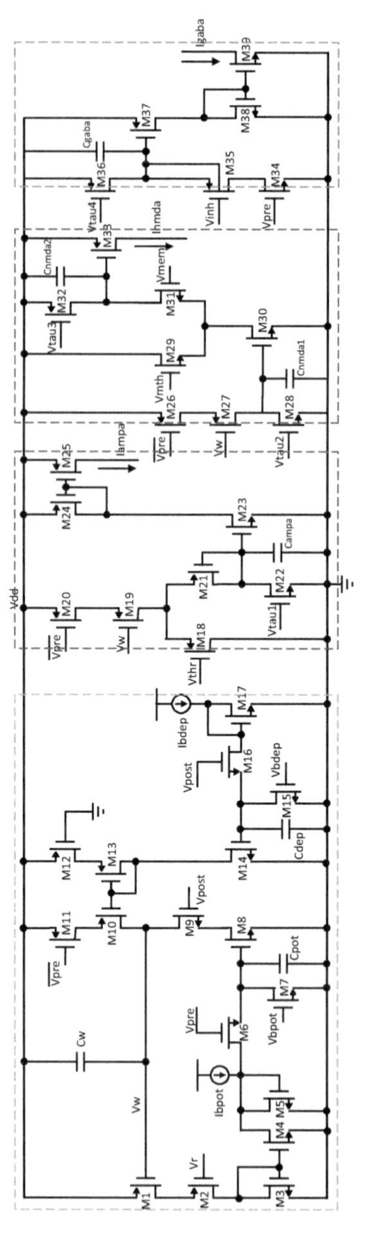

Figure 8.15 The cluster circuit details.

When $V_{mem} > V_{mth}$, the positive feedback drives abruptly the V_{mem} towards the high rail. Meanwhile, the digital output V_{out} of this neuron is turned to "on" state, which activates the reset branch. When V_{mem} is below V_{mth} after the reset leakage, V_{out} recovers to "off" state, inducing an abrupt decrease of V_{mem}. In this way, the action potentials are produced through charging and discharging processes of the membrane voltage.

8.4.4.2 Results

In Figure 8.16(a), AMPA currents are introduced at the onset time of 5 ms, and NMDA currents are injected at different times. Two cases need to be discussed individually. In the first case, where AMPA stimuli precedes NMDA, the excitatory function of NMDA receptors can be demonstrated by the increase of V_{mem} (Δt = 2, 5, 8 ms). As NMDA stimuli approaches AMPA stimuli, larger V_{mem} is detected by NMDA synapse, which gives a greater voltage amplification.

However, if delivered in reversed sequence (Δt = –1 ms), no modification is observed. This can be principally explained by the cooperation mechanism of those two receptors, i.e., AMPA receptors usually act as preliminary depolarization of post neurons by inducing a small amount of ions (Na^+) into cells. When depolarization threshold is surpassed, NMDA receptors are activated, which allows substantial incursion of ions (both Na^+ and Ca^{2+}) and bigger electrical stimuli are produced. Thus, it is implied that NMDA

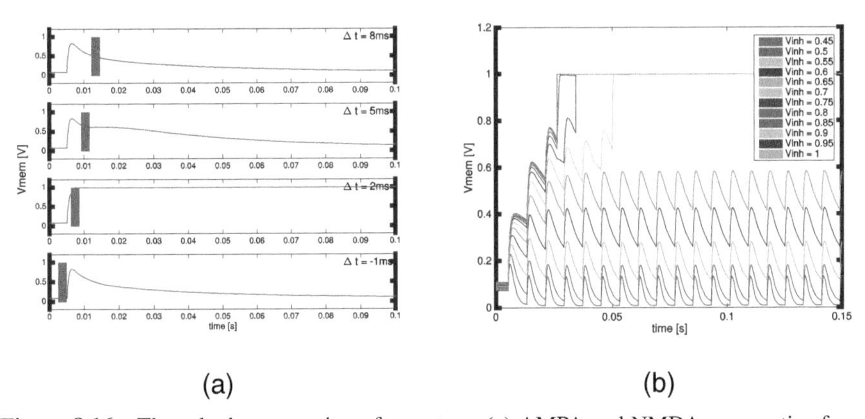

Figure 8.16 The role demonstration of receptors. (a) AMPA and NMDA cooperative function. The AMPA current are induced at an onset time of 5 ms while that for NMDA receptors varies (labeled with red vertical lines). Δt represents the interval between NMDA and AMPA activations, ranging from –1 to 8 ms; (b) The balance function of GABAa receptors.

receptors are not self-initiated. However, once activated, the NMDA receptor acts as a major contribution to electrical signal transmission in the neuron system.

In Figure 8.16(b), the contribution of inhibitory synapses to synapse integration is identified. Various levels of inhibition are applied to the system, while the setting of excitatory synapses are maintained. When inhibition behavior is larger than a certain level ($V_{inh} \leq 0.65$ V), neuron system operates normally. Conversely, if the inhibition level decreases, excitation prevails, driving membrane state to the upper boundary, and consequently information may be lost during this process.

The joint functionality of receptors can also be indicated in a spike pattern. Input spikes at a rate of 100 Hz where prespikes precede post-spikes (for 1 ms) are introduced into synaptic learning circuit, inducing consecutive depression to synaptic weight. In the presence of three receptors (Figure 8.17(a)), 10 membrane spikes are generated. A gradually sparser distribution of the spikes is observed along with the decline of synaptic weight. When synaptic weight reaches lower bound, the network fails to produce any spike trains. In Figure 8.17(b), the function of NMDA receptor is inhibited. Although some post activities occur, the temporal intervals to generate equal number of spikes as Figure 8.17(a) are larger, and the amount of spike clusters is lower due to a lack of long-term dynamics. NMDA receptor acts as a supplement to synaptic excitation. On the other hand, when the function of AMPA receptor is forbidden, no postspike trains are observed (shown in Figure 8.17(c)). This result is coherent with that of biological experiments observed in hippocampal region [21]. Synapse with only NMDA receptors, also called silent synapse, will only transmit information when the postsynaptic neuron is depolarized, caused by synchrony pairing of other synapses with AMPA receptors. Otherwise, a minimal current will be produced. In Figure 8.17(d), the GABAa receptor is blocked. Consequently, the network fails to transmit learning information carried by synapses. Hence, GABAa receptor is essential to create stable signal transmission in SNNs.

In summary, each type of receptor has its distinctive role in the neural system to ensure accurate and stable processing of input signals. Due to the sub-threshold region operation and a high threshold voltage implementation, the receptors have low energy consumption of only 1.92, 3.36, and 1.11 pJ/spike respectively (tested under an input frequency of 100 Hz), achieving a balance between biological complexity and configuration diversity.

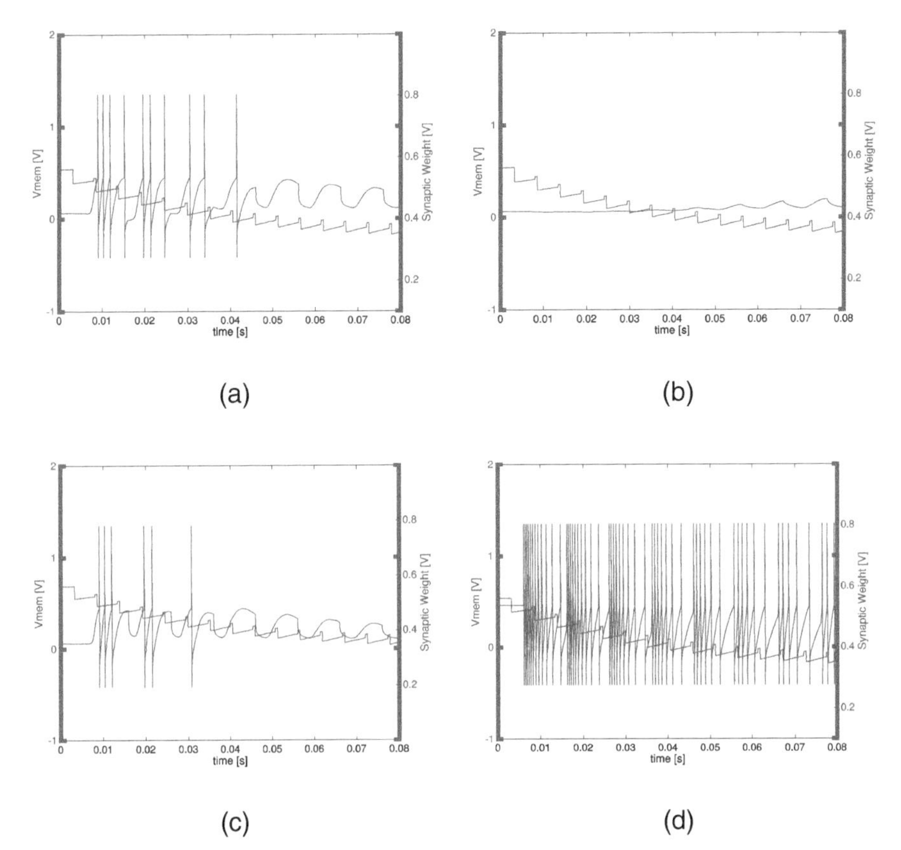

Figure 8.17 The role demonstration of receptors in spike patterns. (a) Spike response of three receptors; (b) spike response without NMDA receptor; (c) spike response without AMPA receptor; and (d) spike response without GABAa receptor.

8.5 Neural Network with Multi-Receptor Synapses

8.5.1 Synchrony Detection Tool: Cross-Correlograms

Sample cross-correlograms are displayed in Figure 8.18. The left plot shows no dependence, while the right one shows strong correlation. It is a visualization of cross-correlation between two spike trains, i.e., the similarity of two series as a function of the temporal displacement of one relative to the other. In this cross-correlogram, the temporal differences between every single pair of spikes are summed for certain temporal bin. A peak present in the cross-correlogram indicates a correlated relation at this certain temporal

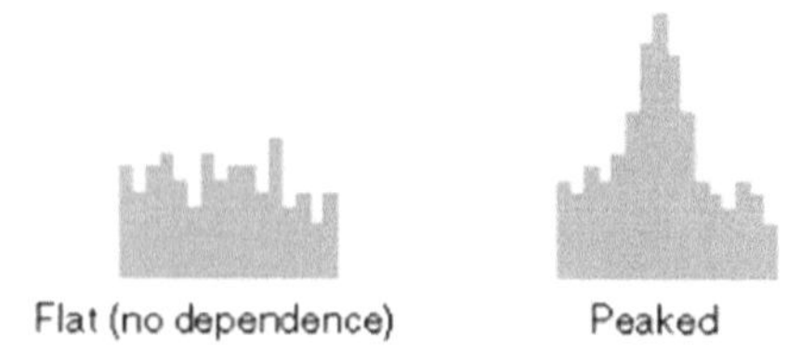

Figure 8.18 Sample cross-correlograms.

bin between target spike groups. For discrete signals, the cross-correlation is defined as [23]:

$$(f * g)[n] \stackrel{\text{def}}{=} \sum_{m=-\infty}^{\infty} f^*[m]\, g[m+n] \tag{8.7}$$

where f^* denotes the complex conjugate of f and n is the displacement bins, which correspond to the temporal difference between two target spikes.

The magnitude of the correlation in the cross-correlogram indicates the causality level between sequential neural units. The cross-correlogram is an efficient way to characterize the computation ability of artificial neural network, and especially the SNN where the weight update information is encoded into the temporal difference between spikes. When valid learning updates occur, the spikes generated are more closely distributed (regardless of the transmission delay), leading to a higher cross-correlation between two spike trains.

8.5.2 Environment Settings

In this section, a two-layer recurrent network illustrated in Figure 8.19 is utilized to explore the parallel and hierarchical synchrony detection and amplification function of the multi-compartment synapse. The same simulation environment (network configuration, input and noise patterns) is applied to three receptor settings: multi-receptor, AMPA-receptor, and NMDA receptor, as shown in the dashed blocks in Figure 8.19. Each setting consists of three neural clusters, which includes one classic I&F neuron and four functional synapses with either multiple- or single-receptor implementations as described in Section 8.3.4. A correlated Poisson distributed spike train C1 is introduced to the first two synapses out of four of every cluster, while another correlation C2 is added between two C1, which adds an additional

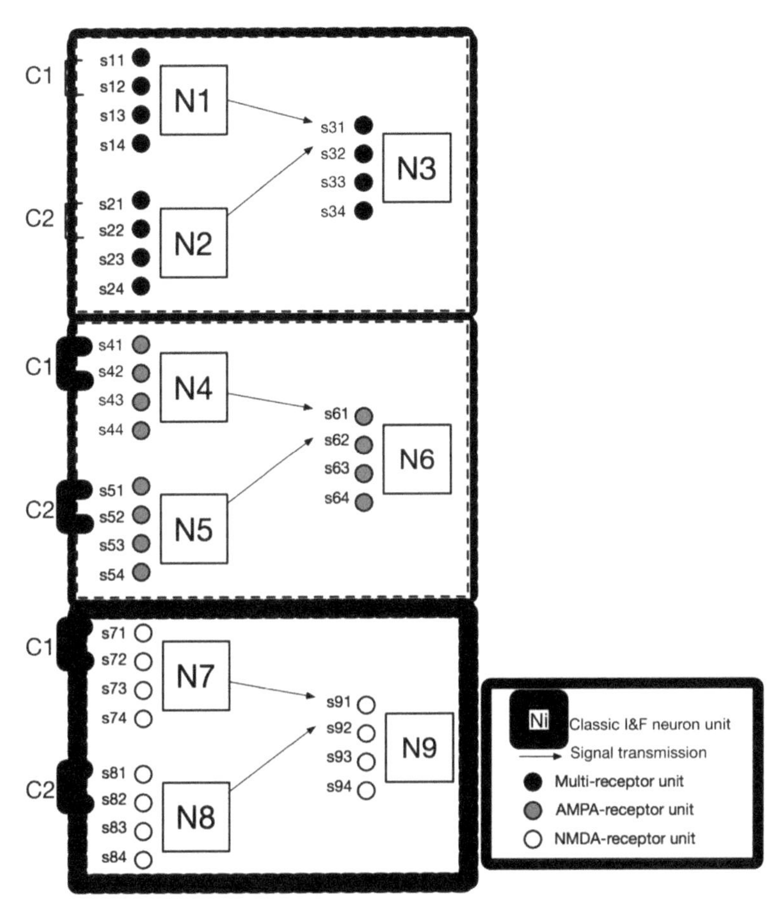

Figure 8.19 Top-level diagram of the two-layer recurrent testing network. The symbol interpretations are listed in the box on the right. Each neuron Ni are connected with four synapses si1-si4, forming a cluster unit discussed in Section 8.3.4. Clusters belonging to different dashed blocks include different types of synaptic receptor configurations. Input C1 is a Poisson distributed spike train of 40 Hz. Input C2 is correlated with C1, and this correlation can be in any correlation form. Here, a delay of 2 ms is used. The rest of the synapses receive Poisson distributed spike trains of 15 Hz.

correlation between N1 and N2 (also N4 and N5, N7, and N8). The rest of the synapses obtain Poisson distributed spike trains as noise. Highly correlated spike trains are more likely to coincide in the defined learning window of STDP learning, which will cause more valid weight update events.

8.5.3 Input Patterns

Among the numerous sources of noises existing in neuron network, the synaptic noise is the main contribution [24]. The synapses release packets of neurotransmitters at the axon terminals depending on the history of firing of both the pre- and postsynaptic neurons; on the other hand, the learning induces long-term effect on the postsynaptic neurons, which would transmit the resultant spike trains to every spatial location they can reach and further form a recurrent network. This non-unidirectional transmission may pass on those spikes to a cell as noises. The summation of thousands of synaptic inputs of one neuron forms irregular fluctuations on the neural response, ranging from completely random Poisson inputs to periodic inputs [24]. The probability of k events in an interval P(x) is given by:

$$P(x) = \frac{e^{-\lambda}\lambda^{x}}{x!} \tag{8.8}$$

where λ is the average number of events per interval and k ranges from 0 to the temporal interval concerned. The Poisson distributions with five different λ values are illustrated in Figure 8.20.

The correlated input spikes induced by C1 and C2 correlation can have several forms: it can be perfectly simultaneous, or one precedes the other, or one leads the other. An example of cross-correlation with different delay values is given in Figure 8.21(a). Larger delay time decreases the possibility of coincidence between two spike trains within target time range, thus reducing the peak correlation amplitude. In addition, a temporal shift is generated, which is proportional to the delay introduced to C2.

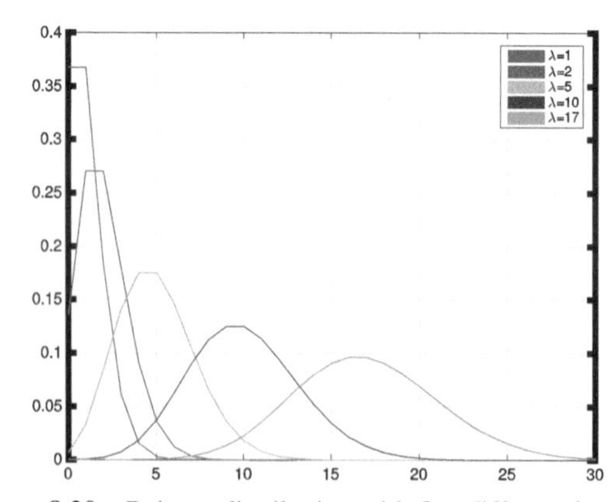

Figure 8.20 Poisson distribution with five different λ values.

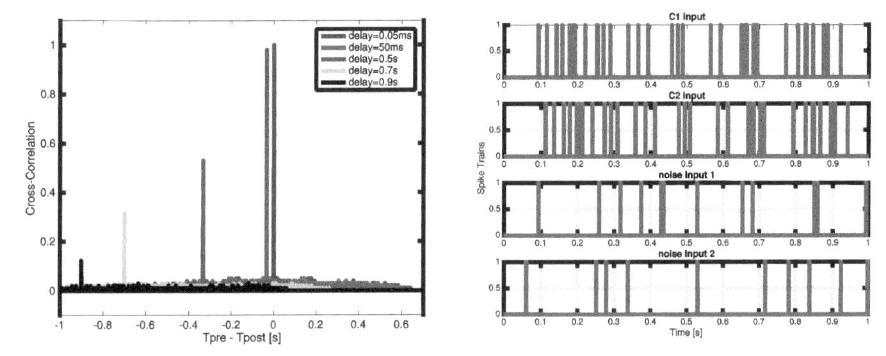

Figure 8.21 (a) The effect of delay of spike trains on cross-correlation. (b) Example input spike trains.

The decay time is set as 2 ms. An example of input signals C1, C2 and two noise signals are displayed in Figure 8.21(b).

8.5.4 Synchrony Detection

Figure 8.22 shows the normalized cross-correlogram results from the two-layer recurrent testing network described above. It compares in parallel the synchrony level of three different receptor configurations: multi-receptor, AMPA-receptor, and NMDA receptor. In every plot, the temporal range of interest is 0.1 s, which covers the range of the learning window reported in [25]. The correlation levels are normalized for better comparison. As observed in Figure 8.21(a), the shift of correlation spikes implies the temporal difference between two spike trains. On the other hand, the amplitude of the correlation peak denotes the correlation level, i.e., the total amount of the coincidence occurring at this target lag point.

The histograms in Figure 8.22(a) and (b) evaluate the cross-correlations between parallel clusters N1, N2 and hierarchical clusters N1, N3. A large level of correlation is observed at close to zero time-point for both cases. This indicates a strong synchrony between both parallel and hierarchical spike trains after synaptic learning process with multi-receptor settings.

In Figure 8.22(c) and (d), the synchrony level is decreased almost by half. In the synapse unit, the advanced STDP learning circuit (see Section 8.2.2), which incorporates both the presynaptic and postsynaptic spike activities to induce learning, presents certain level of synchrony detection [17]. However, the large discrepancy between the two correlation levels does not originate from the learning circuit but from the multi-receptor synapses, since both

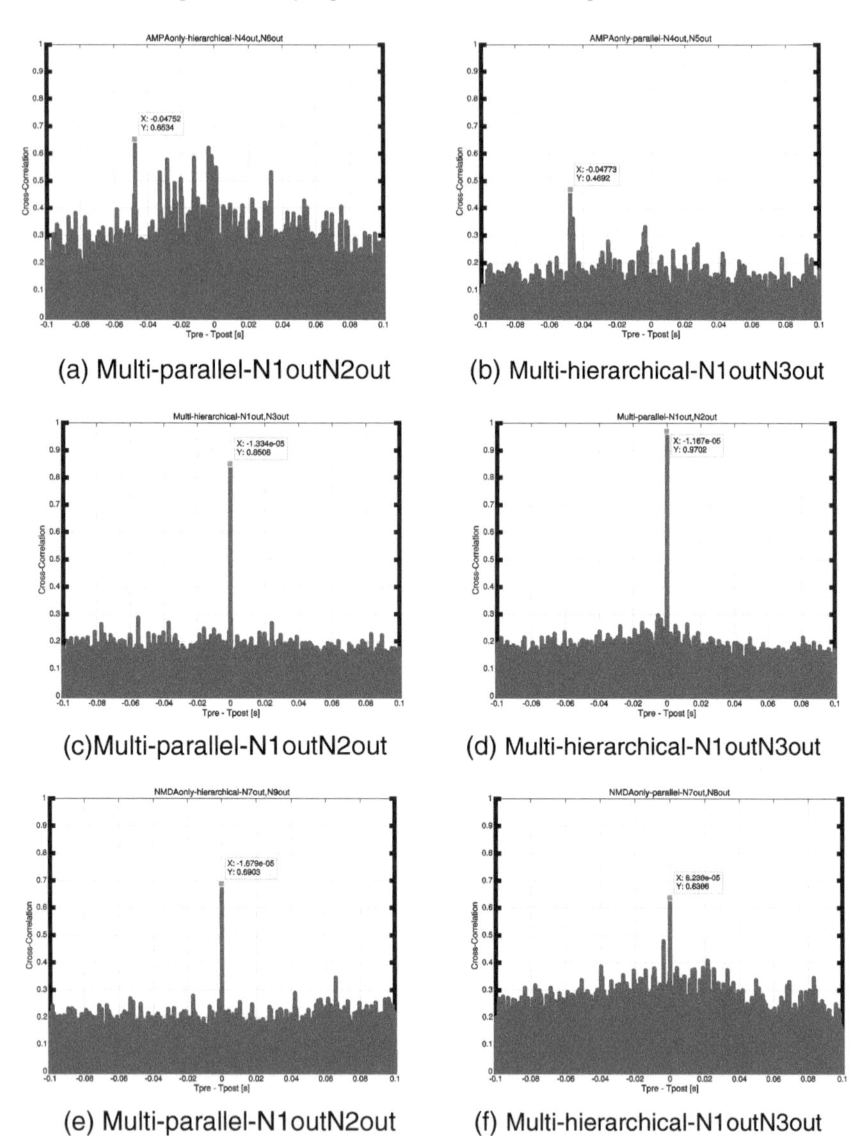

(a) Multi-parallel-N1outN2out (b) Multi-hierarchical-N1outN3out

(c)Multi-parallel-N1outN2out (d) Multi-hierarchical-N1outN3out

(e) Multi-parallel-N1outN2out (f) Multi-hierarchical-N1outN3out

Figure 8.22 Normalized cross-correlogram results from the two-layer recurrent testing network. (a)(b), (c)(d), and (e)(f) are the parallel and hierarchical cross-correlation plots of multi-receptor, AMPA-receptor, and NMDA-receptor configurations, respectively. The annotation above each figure correspond to "receptor configuration-correlation source type-neuron numbers". For example, "Multi-parallel-N1outN2out" means the parallel correlation of multi-receptor settings between neuron clusters N1 and N2.

Table 8.4 Normalized cross-correlation comparison

	AMPA-receptor		NMDA-receptor		Multi-receptor	
	Location(s)	Amplitude	Location(s)	Amplitude	Location(s)	Amplitude
Parallel	−0.04773	0.4692	−0.238e-5	0.6386	−1.167e-5	0.9702
Hierarchical	−0.04752	0.6534	−1.679e-5	0.6903	−1.333e-5	0.8508

of the systems have the same learning implementations. Additionally, the background noise is observed as well as several sub-peaks occurring near the origin in the hierarchical relations, which implies a relatively poorer stability performance. Along with the amplitude decay, a peak shift occurs. The delay between the inputs to the clusters is passed through layers, while that of the multi-receptor synapses is mitigated. Finally, Figure 8.22(e) and (f) characterize the synchrony detection function of NMDA-receptor network. Both parallel and hierarchical pairs have similar correlation plots as multi-receptor network but with reduced amplitudes (around 60–70% as that of multi-receptor).

The detection and amplification levels of the cross-correlation function of various spike train pairs using both single and multiple receptors are shown in Table 8.4. The maximum amplification level of multi-receptor configuration is almost 2 times higher than that of single-receptor ones.

In summary, the analog multi-compartment synapse structure is able to detect and amplify the temporal synchrony embedded in the synaptic noise. The maximum amplification level is 2 times larger than that of single-receptor configurations. Moreover, the circuit shows efficient learning ability; the consecutive neural clusters generate almost synchronized output spike patterns in the presence of delay in inputs signals, i.e., it takes shorter time for system with multiple-receptor to achieve synchrony. Analysis indicate that this ability originates from the NMDA-receptor, since NMDA-receptor displays similar correlations, except for a decrement in the amplitude of correlation level. Evidently, AMPA and NMDA receptors have a collaborate relation in inducing efficient synchrony detection for synapse structures.

8.6 Conclusions

In this chapter, we propose a current-based neuromorphic synapse architecture for SNN, which incorporates the structures of the weight-dependent learning rule and multiple receptors, namely AMPA, NMDA, and GABAa, and thus provides distinctive temporal dynamics of each type of receptors in one synapse design. Improved synchrony detection and amplification ability is demonstrated through cross-correlation study. The synaptic design implemented in TSMC 65 nm CMOS technology consumes 1.92, 3.36, 1.11, and 35.22 pJ per synaptic event for AMPA, NMDA, GABAa receptors, and the advanced STDP learning circuit, respectively.

References

[1] G. L. Shaw, Donald Hebb: The Organization of Behavior. Berlin, Heidelberg: Springer Berlin Heidelberg, pp. 231–233 1986 [Online]. Available at: http://dx.doi.org/10.1007/978-3-642-70911-1-15

[2] K. Gerrow and A. Triller, "Synaptic stability and plasticity in a floating world," Current Opinion in Neurobiology, vol. 20, no. 5, pp. 631–639, 2010, neuronal and glial cell biology New technologies. [Online]. Available at: http://www.sciencedirect.com/science/article/pii/S0959438810001078

[3] W. Gerstner, W. M. Kistler, R. Naud, and L. Paninski, Neuronal dynamics: From single neurons to networks and models of cognition. Cambridge University Press, 2014.

[4] D. Y. Kim, S. H. Kim, H. B. Choi, C.-k. Min, and B. J. Gwag, "High Abundance of GLUR1 mRNA and Reduced Q/R Editing of $GLUR_2$ mRNA in Individual NADPH-Diaphorase Neurons," *Molecular and Cellular Neuroscience*, vol. 17, no. 6, pp. 1025–1033, 2001.

[5] S. R. Platt, "The role of glutamate in central nervous system health and disease–a review," *The Veterinary Journal*, vol. 173, no. 2, pp. 278–286, 2007.

[6] A. M. VanDongen, Biology of the NMDA Receptor. CRC Press, 2008.

[7] A. Destexhe, Z. F. Mainen, and T. J. Sejnowski, "Kinetic models of synaptic transmission," *Methods in neuronal modeling*, vol. 2, pp. 1–25, 1998.

[8] S. H. Wu, C. L. Ma, and J. B. Kelly, "Contribution of ampa, nmda, and gabaa receptors to temporal pattern of postsynaptic responses in the inferior colliculus of the rat," *Journal of Neuroscience*, vol. 24, no. 19, pp. 4625–4634, 2004.

[9] E. De Schutter and I. ebrary, Computational Modeling Methods for Neuroscientists. MIT Press, vol. Computational neuroscience series, 2009.

[10] A. son, M. Diesmann, and W. Gerstner, "Phenomenological models of synaptic plasticity based on spike timing," *Biological cybernetics*, vol. 98, no. 6, pp. 459–478, 2008.

[11] G.-q. Bi and M.-m. Poo, "Synaptic modifications in cultured hippocampal neu- rons: dependence on spike timing, synaptic strength, and postsynaptic cell type," *Journal of neuroscience*, vol. 18, no. 24, pp. 10,464–10,472, 1998.

[12] S. Song, K. D. Miller, and L. F. Abbott, "Competitive hebbian learning through spike-timing-dependent synaptic plasticity," *Nat Neurosci*, vol. 3, no. 9, pp. 919–926, Sep 2000.

[13] J. Rubin, D. D. Lee, and H. Sompolinsky, "Equilibrium properties of temporally asymmetric hebbian plasticity," Phys. Rev. Lett., vol. 86, pp. 364–367, Jan 2001. [Online]. Available at: https://link.aps.org/doi/10.1103/PhysRevLett.86.364

[14] R. Gtig, R. Aharonov, S. Rotter, and H. Sompolinsky, "Learning input correlations through nonlinear temporally asymmetric hebbian plasticity," *Journal of Neuroscience*, vol. 23, no. 9, pp. 3697–3714, 2003. [Online]. Available at: http://www.jneurosci.org/content/23/9/3697

[15] J.-P. Pfister and W. Gerstner, "Triplets of spikes in a model of spike timing dependent plasticity," *Journal of Neuroscience*, vol. 26, no. 38, pp. 9673–9682, 2006.

[16] G. Indiveri, E. Chicca, and R. Douglas, "A vlsi array of low-power spiking neurons and bistable synapses with spike-timing dependent plasticity," *IEEE transactions on neural networks*, vol. 17, no. 1, pp. 211–221, 2006.

[17] A. Bofill-i Petit and A. F. Murray, "Synchrony detection and amplification by silicon neurons with stdp synapses," *IEEE Transactions on Neural Networks*, vol. 15, no. 5, pp. 1296–1304, 2004.

[18] M. R. Azghadi, S. Al-Sarawi, D. Abbott, and N. Iannella, "A neuromorphic vlsi design for spike timing and rate based synaptic plasticity," *Neural Networks*, vol. 45, pp. 70–82, 2013.

[19] C. Bartolozzi and G. Indiveri, "Synaptic dynamics in analog vlsi," *Neural computation*, vol. 19, no. 10, pp. 2581–2603, 2007.

[20] C. Mead and M. Ismail, Analog VLSI implementation of neural systems. Springer Science & Business Media, vol. 80, 2012.

[21] D. Liao, N. A. Hessler, and R. Malinow, "Activation of postsynaptically silent synapses during pairing-induced ltp in ca1 region of hippocampal slice," *Nature*, vol. 375, no. 6530, p. 400, 1995.

[22] A. Destexhe, Z. F. Mainen, and T. J. Sejnowski, "Kinetic models of synaptic transmission," *Methods in neuronal modeling*, vol. 2, pp. 1–25, 1998.

[23] D. J. Amit and S. Fusi, "Learning in neural networks with material synapses," *Neural Computation*, vol. 6, no. 5, pp. 957–982, 1994.
[24] N. Brunel, F. S. Chance, N. Fourcaud, and L. Abbott, "Effects of synaptic noise and filtering on the frequency response of spiking neurons," Physical Review Letters, vol. 86, no. 10, p. 2186, 2001.
[25] G.-q. Bi and M.-m. Poo, "Synaptic modification by correlated activity: Hebb's postulate revisited," *Annual review of neuroscience*, vol. 24, no. 1, pp. 139–166, 2001.
[26] S. H. Wu, C. L. Ma, and J. B. Kelly, "Contribution of ampa, nmda, and gabaa receptors to temporal pattern of postsynaptic responses in the inferior colliculus of the rat," *Journal of Neuroscience*, vol. 24, no. 19, pp. 4625–4634, 2004. [Online]. Available at: http://www.jneurosci.org/content/24/19/4625

9

Conclusion and Future Work

Amir Zjajo and Rene van Leuken

Delft University of Technology, Delft, The Netherlands

9.1 Summary of the Results

Current neuron simulators, which are precise enough to simulate neurons in a biophysically meaningful way, are limited in the amount of neurons to be placed on the chip, the interconnect between the neurons, run-time configurability and the re-synthesis of the system. In this book, a system is proposed that is able to bridge the gap between biophysical accuracy and large numbers of cells (19,200 cells for neighbour connection mode, and over 3000 cells in normal connection mode). The cells are grouped around a shared memory in clusters to allow for instantaneous communication. Clusters that are close communicate using only one hop in the network; clusters that are further away communicate less frequently and, consequently, the penalty for taking multiple hops is less severe. An added advantage is that the system can be extended over multiple chips without significant performance penalty. This combination of clusters and a tree topology network-on-chip allows for almost linear scaling of the system. To provide run-time configurability, a tree-based communication bus is used, which enables the user to configure the connectivity between cells and change the parameters of the calculations. As a result, re-synthesising the whole system just to experiment with a different connectivity between cells is not required. The user has to enter the amount of neurons in the system as well as the desired connectivity scheme. From this information, all required routing tables and topologies are automatically generated, even for multi-chip systems. Porting the network to the FPGA yields at least several thousands of simulation speed-ups in comparison with SystemC simulation, with negligible loss of accuracy.

The main contributions of the proposed system are summarised as follows:

- *Close to linear growth in communication cost.* Simulations show that the system is able to handle 19,200 cells for neighbour connection mode and over 3000 cells in normal connection mode. The constraints used for this result are 50 μs per simulation step at 100 MHz clock frequency (5000 cycles). The cell calculation latency is 528 cycles. Faster cell calculation latencies improve the capabilities of the system.
- *Extendable system that can cover multiple chips.* The current state-of-the-art system simulates 96 cells. By using the multi-chip capabilities of the proposed system, four to eight times the amount of cells are possible. To connect four FPGAs, only moderate inter-FPGA link requirements need to be fulfilled. For eight FPGA systems, a high link speed is required.
- *The proposed system is limited by the calculation speed of the PhC.* The proposed system scales with improvements to the calculation speed as well as improvements to the FPGA size.
- *Simulation environment to test system configurations.* Before implementing a specific configuration, the simulator can be used to determine the performance of the new system. The simulator incorporates all latencies, such as calculation latencies and on- and off-chip communication latencies.
- *Automatic structure and connectivity table generation.* The system generates all necessary routing tables and structures automatically when provided with the desired specifications, such as cell amount and desired connectivity scheme.
- *Higher biophysical accuracy compared to the state of the art.* The resulting system with massive amounts of cells can be used in a variety of research areas, such as artificial intelligence or driving assistance.

The neural network configuration parameters have an effect on both the latency of the design and the amount of required resources; however, each with a different degree. In Chapter 3, first the effect of changing the time-sharing factor is inspected, then the number of physical cells, the number of cluster controllers and finally the ExpC factor. By changing the time-sharing factor, multiple neuron responses are computed over the same hardware; this has resulted in a large linear effect on the overall latency of the design; however, critical resources are not affected. By changing the number of physical cells connected to a cluster controller, the size of the routing network does not increase and the cluster controller is better utilised; however, there is only one central memory over multiple physical cells and they have the possibility of blocking each other. Changing the number of cluster controllers in the design has a slight effect on the latency due to a communication delay between controllers, although it has a larger effect on the usage of critical resources than increasing the number of physical cells. Finally, by changing the ExpC factor, the design is tweaked to find a satisfying balance between critical resources and speed.

Real-time reconfigurable learning neuron networks are not only limited by run-time configurability, the re-synthesis of the system and the interconnect between the neurons, but also mainly by the amount of neurons that can be placed on the chip. In Chapter 4, we implement several models of the spiking neurons with axon conduction delays and spike timing-dependent plasticity in a real-time data-flow learning network. The system implemented on the Virtex 7-XC7VX55 device can accommodate 1188 Hodgkin–Huxley-type neuron cells and approximately 2800 and 3000 Izhikevich and integrate-and-fire-type cells, respectively. A tree-based communication bus is utilised since it offers run-time configurability, e.g. the user-enabled configuration of the connectivity between cells and the adaptability of calculation parameters. Consequently, the system does not need to be re-synthesised just to experiment with a different connectivity between cells. The cells are grouped around a shared memory in clusters to allow instantaneous communication.

In Chapter 5, the multipath ring (MPR) topology is proposed as an efficient dataflow architecture for neuron-to-neuron communication in a clustered spiking neuron network simulator to be implemented on FPGA. When compared with the two-dimensional torus, an architecture commonly employed for multi-core platforms, improved throughput capacity was found, along with the possibility of traffic shaping, or redirecting through specific links. By configuring MPR parameters such as skipping distance, the topology can be adapted for specific network sizes and to maintain very efficient traffic balance. The MPR is capable of handling different neuron communication schemes as well, which makes it suitable for supporting the continuous changes in traffic patterns due to learning processes in the neural network.

As a method for characterising the energy efficiency of the multipath ring network topology, the energy-delay product was estimated and compared with other network types, such as the low-power inverse-clustering with mesh. The observed values indicate an energy-efficient topology that has the added benefit of low network diameter and physical implementation simplicity.

Strain on the network was reduced using two physically separate communication layers, one for the neuron-to-neuron data, for which the MPR was proposed, and the other for configuration and input/output data. For the former, a binary-tree topology was adapted, with a channel routing protocol for downstream traffic. In the case of upstream traffic, no routing decisions have to be made apart from arbitration. Overall, the routers can be kept simple and do not require routing tables.

In Chapter 6, a hierarchical dataflow architecture is proposed that is capable of bridging the gap between biophysical accuracy and large numbers (50k) of cells.

In Chapter 7, a novel ultra-low-power ECG beat-type classification system for arrhythmia detection has been proposed. In the system, an adaptive ECG interval extraction is utilised for feature extraction, and correlation matrix for unsupervised feature selection. The classification of the resulting features is performed with neuromorphic liquid-state machine. The proposed system has characteristics of the global classifier trained in an unsupervised manner, which does not require labelled data for training, and consequently, it can be adapted to the individual patient. For compatibility with wearable devices, the classification system requires only one ECG lead. The system as a whole is estimated to consume less than 1 μW and has an overall clustering accuracy of 95.5%. The performances achieved are comparable to those reported in the literature for fully automated, multi-lead algorithms.

In Chapter 8, a novel synapse structure incorporating an advanced STDP learning algorithm and multi-compartment synapse has been proposed. Three synaptic learning circuits are proposed: the classic STDP, the advanced STDP and the triplet STDP circuits. The classic STDP circuit does not exhibit basic exponential and real-time dynamics, although the area and power features are the best among three. The triplet STDP is able to capture diverse experimental observations obtained from real neural system. The corresponding area and power consumptions are almost doubled compared with the advanced STDP circuit. The advanced STDP circuit achieves a great trade-off between biological fidelity and resource consumptions. A wide temporal range of synaptic learning windows up to 100 ms is possible. Additionally, a weight dependence feature expands its biological properties. The energy consumed per spike event is merely 35.22 pJ.

The multi-compartment synapse design gives biologically accurate modelling of chemical synapse, which increases the computation ability of synapses. Each type of receptor forming the synapse structure plays a distinctive role in the neural system to ensure accurate and stable processing of input signals. More importantly, the cross-correlated input patterns can be well detected and amplified through layers of multi-compartment synapse structure. Better synchrony indicates a higher efficiency in signal processing and better computation ability in recurrent neural systems. Moreover, the receptors have extremely low energy consumption of only 1.92, 3.36, and 1.11 pJ/spike, representing a good balance between biological complexity and configuration diversity.

9.2 Recommendations and Future Work

A real-time, reconfigurable, multi-chip system, which allows large-scale, biophysically accurate neuron simulation, should exhibit several characteristics:

Implementation on FPGA: First and foremost, the system should be implemented on an FPGA to prove that the simulation results are correct and hold true for hardware implementations. After the completing a cycle-accurate SystemC, mainly transmission work is necessary to generate synthesisable SystemC. Alternatively, the system can be rewritten in any hardware description language (HDL).

Analog Simulation: SystemC AMS can be used to simulate the analog part of the system, e.g., analog-to-digital and digital-to-analog converters to simulate the real life behaviour of the converters and feed the system with actual data.

Calculation Performance: Reducing the time required for calculations increases the cluster sizes, which results in higher communication localisation and thus better interconnect performance, as less communication leaves a cluster.

Router Memory Usage: Depending on the type of FPGA, it can be advantageous to explore reducing the size of the routing tables, as well as the various buffers inside the routers to allow for further scaling. Currently, on Virtex 7 and on Spartan 6 FPGA, the routers are small enough to support systems with several thousand cells across multiple FPGA.

Second Clock Domain: As the interconnect can be run at higher clock frequencies compared to the PhC, the communication costs can further be lowered by introducing a second faster clock for the interconnect. The clock domains can either be split at the cluster boundaries, which are already implemented as *FIFO*, thus making clock syncing easier or at the PhC domains, which is possible as the shared memory in the clusters is implemented as true dual-port memory. The latter approach requires more synchronisation effort, and currently, would not provide any performance benefit over the former approach.

Multi-chip Topology: Topologies for the multi-chip interconnect is required, which is fast and also implementable in real hardware to allow for larger, tightly packed systems.

Automatic Learning: The process of constructing and deconstructing connections between cells can be automated to simulate learning processes. Such algorithms can be implemented either by using the already implemented control bus or by using hardware as new modules of the system. For the first approach, a way to read information from the system, is required to receive statistics about the different cell connections that can be used by the algorithms to decide about new connections or to destroy connections. The second approach might be unfeasible due to hardware costs.

Event-based Approach: To increase the efficiency of the system, the system should only react on events instead of always producing new data, i.e. each communication packet is delivered despite its content (also without information).

Optimisation of Cell Placements: Optimisation algorithms should be used to optimise the distribution of cells on the clusters with a goal to achieve a cell distribution that reduces communication over cluster of chip boundaries.

Priority Packets: In certain situations, it could happen that packets that have to travel across chip boundaries are calculated too late in an iteration cycle. Packets that have to travel further should be generated first by the clusters, which would allow for more communication to be hidden behind the calculations.

Index

About the Editors

Amir Zjajo received the M.Sc. and DIC degrees from the Imperial College London, London, U.K., in 2000 and the PhD. degree from Eindhoven University of Technology, Eindhoven, The Netherlands in 2010, all in electrical engineering. In 2000, he joined Philips Research Laboratories as a member of the research staff in the Mixed-Signal Circuits and Systems Group. From 2006 until 2009, he was with Corporate Research of NXP Semiconductors as a Senior Research Scientist. In 2009, he joined Delft University of Technology, Delft, The Netherlands, as a Faculty Member with the Circuits and Systems Group. In 2018, he co-founded Innatera Nanosystems B.V. to commercialize bionic signal processing technology.

Dr. Zjajo has published more than 80 papers in referenced journals and conference proceedings, and holds more than 10 US patents or patent pending. He is author of the books *Brain-Machine Interface: Circuits and Systems* (Springer, 2016), *Low-Voltage High-Resolution A/D Converters: Design, Test and Calibration* (Springer, 2011, Chinese translation, China Machine Press, 2015), and *Stochastic Process Variations in Deep-Submicron CMOS: Circuits and Algorithms* (Springer, 2013). He served as a member of Technical Program Committee of IEEE International Symposium on Quality Electronic Design, IEEE Design, Automation and Test in Europe Conference, IEEE International Symposium on Circuits and Systems, IEEE International Symposium on VLSI, IEEE International Symposium on Nanoelectronic and Information Systems, and IEEE International Conference on Embedded Computer Systems.

His research interests include energy-efficient digital/mixed-signal circuit and system design for biomedical and mobile applications, on-chip machine learning and inference, sensor fusion, and bionic electronic circuits for autonomous cognitive systems.

Rene van Leuken received the M.Sc. and Ph.D. degrees in electrical engineering from the Delft University of Technology, Delft, The Netherlands, in 1983 and 1988, respectively. He is currently a Professor with the Circuit and Systems Group, Faculty of Electrical Engineering, Mathematics and Computer Science, Delft University of Technology (TU Delft), The Netherlands.

He has authored or coauthored papers in all major journals, conferences and workshops proceedings, and has received several best paper awards over the years. His research interests include high-level digital system design, system design optimization, VLSI design, and high performance compute (DSP) engines. His major research activity is neuromorphic computing.

Dr. van Leuken has been involved in many major research and development projects: ESPRIT, FP6, FP7, JESSI, MEDEA, and recently in ENIAC/CATRENE, and ARTEMIS projects. He is member of the PATMOS steering committee and the DATE Technical Program Committee.